International Environmental Management Benchmarks

Springer

Berlin
Heidelberg
New York
Barcelona
Hong Kong
London
Milan
Paris
Singapore
Tokyo

David M.W.N. Hitchens · Jens Clausen · Klaus Fichter (Eds.)

International Environmental Management Benchmarks

Best Practice Experiences from America, Japan and Europe

With 40 Figures and 16 Tables

 Springer

Editors

Prof. David M.W.N. Hitchens
The Queen's University of Belfast
Department of Economics
Belfast BT7 1NN, Northern Ireland

Dipl.-Ing. Jens Clausen
Institut für ökologische
Wirtschaftsforschung (IÖW) gGmbH
Hausmannstraße 9–10
D-30159 Hannover, Germany

Dipl.-Ökonom Klaus Fichter
Institut für ökologische
Wirtschaftsforschung (IÖW) gGmbH
Potsdamerstraße 105
D-10785 Berlin, Germany

Translator

Gisela Jaeger
Staubstraße 26
CH-8038 Zürich, Switzerland

ISBN 3-540-65296-5 Springer-Verlag Berlin Heidelberg New York

Library of Congress Cataloging-in-Publication Data
International environmental management benchmarks : best practice experiences from America, Japan, and Europe / David M.W.N. Hitchens, Jens Clausen, Klaus Fichter (eds.).
p. cm.
ISBN 3-540-65296-5 (alk. paper)
1. Environmental management--Handbooks, manuals, etc. 2. Environmental management--United States--Handbooks, manuals, etc. 3. Environmental management--Japan--Handbooks, manuals, etc. 4. Environmental management--Europe--Handbooks, manuals, etc. I. Hitchens, D. M. W. N. II. Clausen, Jens, 1958- III. Fichter, Klaus, 1962- GE300 .I6 1999 363.7'056--dc21 98-32086 CIP

© Springer-Verlag Berlin Heidelberg 1999
Printed in Germany

The use of general descriptive names, registered names, trademarks, etc. in this publication does not imply, even in the absence of a specific statement, that such names are exempt from the relevant protective laws and regulations and therefore free for general use.

Cover Design: de'blik, Berlin
Dataconversion: Büro Stasch, Bayreuth

SPIN: 10568107   30/3020 – 5 4 3 2 1 0 – Printed on acid-free paper

Preface

Sustainable development requires that the quality of the environment must be sustained while the economy continues to develop. This is a very broad definition. In fact, not only are there more than 70 different definitions of 'sustainable development' but, moreover, the underlying relationship between environmental damage and economic development is not fully understood. In the face of such uncertainties, how can a company work towards sustainability and contribute to sustainable development? This is the subject of this book.

Today the firm's sphere of influence is wider than that of producer and polluter. It can shape lifestyles and consumer tastes and preferences through publicity, marketing, public relations and the production of new products. The firm can therefore play a major role in sustainability (irrespective of any precise definition of the term) through production and social, economic, and ecological innovation.

The first part of the book is concerned with defining the requirements to be met by the sustainable company and the framework in which it should operate. How can a company contribute to sustainable development? When is a company acting sustainably and when is it not? The two introductory chapters by Klaus Fichter and Jens Clausen/ Maite Mathes from Germany attempt to answer these questions. They define a broad area of responsibility which includes not only economic but also ecological and social factors. Six other articles describe very different elements of the framework for company sustainability. David Hitchens from northern Ireland points out that environmental regulation and compliance are not likely to have a serious effect on company competitiveness even when competing with firms in countries with less stringent environmental regulations. Thomas Dyllick from Switzerland presents ecological strategies to market development. Hugo Kuijjer from the Netherlands describes the actions taken by the Dutch government to develop environmental management systems in industry and the progress made in the first 6 years of a comprehensive programme. Michal Aucott from the United States examines a regulation to foster the public right to know about toxic releases. Tron Kleivane from Norway, convenor of the ISO committee for EPE development, assesses the chances of the implementation of environmental performance measurement at the company level. Inge Schuhmacher and colleagues from Switzerland developed an eco-efficiency investment fund in one of the biggest banks of the world – a resource for the eco-efficient company.

The second part of the book focuses on best practice environmental management. Martin Houldin from UK provides insight into auditing techniques. José Lutzenberger, the former environmental minister for Brazil, applies ideas from industrial ecology to a Third World context. Tomo Shibamiya shows how the Japanese involve employees in an environmental management system. Frans Oosterhuis from the Netherlands and Gerd Scholl and Susanne Nisius from Germany discuss developments in environmental product policy. Klaus Fichter and colleagues and Margaret Pierce from the USA describe internal and external considerations involved in environmental disclosure. Matteo Bartolomeo and Federica Ranghieri were among the first to carry out an environmental benchmarking study with members of an entire industrial sector. Helen Howes and colleagues from Canada have developed an approach to full cost accounting. Materials flow management is described by Kathrin Ankele from Germany. The last German article was contributed by Frank Ebinger and colleagues who evaluated two products of Hoechst Chemicals in the light of sustainable product production and use.

Many of the papers in the collection were originally presented at an international conference on environmental management "Steps Towards the Sustainable Company", held in 1996 by the Ecological Economics Research Institute in Berlin [Institut für ökologische Wirtschaftsforschung (IÖW) gGmbH] in cooperation with the United Nations Environment Programme (UNEP – Industry and Environment), the International Network for Environmental Management (INEM), and other national organizations, and supported by the German Federal Foundation for Environment (Deutsche Bundesstiftung Umwelt). The contributors from Japan, Canada, USA, Brazil, and seven European countries discussed a range of instruments and approaches to environmental management, successfully developed and practised in their own country but hardly known about outside the national boundary. IÖW, the conference organisers were strongly encouraged to disseminate the findings as a step towards the development of the sustainable company.

We would like to thank the authors for their contributions, the staff of the Springer Publishing House in Heidelberg, and Jennifer Rackles in Berlin, who greatly supported us in compiling Part C of this book: "International Organizations and Networks Promoting Environmental Management". Last, but not least, we wish to thank Gisela Jaeger for translating a number of German texts, and Bill Thompson for help with editing a number of chapters.

David M.W.N. Hitchens, Jens Clausen and Klaus Fichter

Queen's University, Belfast
Institut für ökologische Wirtschaftsforschung, Hanover and Berlin

Contents

A

Requirements and Framework for the Sustainable Company

1 Steps Towards the Sustainable Company – Requirements and Strategic Starting Points

Klaus Fichter

Companies are for several reasons an important factor in the process of realizing a sustainable, environment-friendly development. Especially big companies act globally today and represent social centres of power that significantly shape the use of resources and the release of substances and energies, not only by their production but also by their influence on life-styles and consumption patterns. Due to their important role in society, companies, their activities, and the impacts of their activities have to be a matter of the debate about sustainable development. On the one hand, companies cause problems through their shortsighted profit-orientation, through job cutbacks, and through their ruthless exploitation of nature. On the other hand, they may also solve problems by potentially developing social, economic, and ecological innovations.

What can be a company's contribution to sustainable development, and when does a company act in a sustainable way or not? Lasting answers to these questions cannot be attained by 'imposing' governmental targets about environmental quality and environmental action on companies. Lasting answers will have to start with defining the role of companies in today's world and analyzing at first, independent of normative targets, the framework conditions and influences under which companies act today.

Companies Creating Additional Value for Various Stakeholders

As a means to adequately ascertain internal and external requirements that affect a company and ultimately determine its activities, applied economics and the science of business management have developed the 'stakeholder' concept. The stakeholder concept distinguishes internal and external stakeholders. Internal stakeholders are owners, management, and employees. Among the external stakeholders we find, for example, external investors, suppliers, customers, neighbours, environmental associations, the press, and authorities. The stakeholder concept makes clear that there is a multitude of persons and groups which, on the one hand, are affected by a company's activities, and on whose services the company has to rely, on the other hand.

This does away with the assumption underlying classical economics and the related theories of applied economics that there is a harmony between the pursuit of one's individual private economic objectives and the goals of the whole society, a harmony created by the 'invisible hand' of the market. The stakeholder concept rather focuses on the fact that different groups pursue different and quite conflicting goals and raise corresponding claims.

Thus, non-market relations and goals are additionally taken into account, all impacts of a company are assessed as of business relevance – of direct relevance for the company itself. "We can define as external effects those mutual influences on the part of economic subjects which are not registered and evaluated by the market" (Wicke 1989, p. 43).

A company's success is then no longer a consequence of its market success alone, it also depends on other, non-market factors. If, for instance, the company finds out that a stakeholder group, due to external effects like noise emission and pollution, no longer provides certain services important for the firm's existence (here: tolerating these conditions), the company has to deal with these effects irrespective of official regulations, as it does with factors relevant for its market position.

Doing away with the harmony premise means furthermore to abandon the idea that a company legitimizes its activities simply by producing, with a sufficient profit, goods that are demanded on the market. This view underlay classical economics. The theory maintained that the entrepreneur contributes to maximizing the overall social utility by maximizing his individual utility – that is, according to the classical, purely monetary perspective: his profit. Renouncing profit maximization also means – reverting the logical conclusion – reducing social benefit. "Private property is functionally legitimized in this model by its significance for the macroeconomic optimum" (Ulrich 1980, p. 32).

"With the premises of such a metaphysics of the market, the unconditional strive for profit maximization ... does not only seem to be a legitimate right but moreover a moral duty of the entrepreneur in the interest of general welfare" (Ulrich/ Fluri 1995, p. 59).

This simple, one-dimensional legitimization has to be abandoned when the focus is now also on external effects (because then contributions of actors concerned are assessed as equally relevant for the company) and when overall social utility is no longer defined only in terms of material values. Instead, legitimization then also presupposes consideration of interests of other relevant stakeholders. So the stakeholder concept implies a completely different understanding of an enterprise in terms of its function and legitimization in today's world. According to this concept, a company can be defined as a "multifunctional and consequently pluralistically legitimized added-value-creating unit which fulfils socio-economic functions for various stakeholders ... " (Ulrich/Fluri 1995, p. 60).

So, the basic function of an enterprise consists, in the first place, in creating economic values by providing products and services against payment for buyers. The created monetary values which the company keeps (value added) then serve to realize further functions, like creating an income for the employees, paying interests on capital, paying taxes (to finance public expenditures), or taking over social and cultural tasks (social security payments, sponsoring etc.). Moreover, the company has to produce its products and services in such a way that legal provisions are met and claims raised on behalf of superior social interests are considered.

If we describe a company as a multifunctional unit creating additional value, this does not mean, however, that the firm would, could, or means to fulfil all claims raised. On the one hand, these demands result in sometimes considerably conflicting goals, and on the other hand, the recently prevailing discussions about 'shareholder values' show that possibilities to enforce claims still differ widely for individual social groups of stakeholders. This fact is obviously of great relevance for the issue of sustainable development.

Influences on a Company's Activities – Theory of External Interaction Systems

How a company can contribute to sustainable development or what steps it can make towards sustainable business is, on the one hand, a question of framework conditions and external influences constraining its activities. On the other hand, however, this also depends on the use a company makes of its available scope of action and on its efforts to extend it. We will, at first, look at the framework conditions and external influences to which companies are subject in their activities. The theory of external interaction systems provides essential insights here. It offers a comprehensive theoretical basis for determining external influencing factors which are of help in fostering a more sustainable way of companies' economic activities. Then we will discuss the question how companies can use and extend their scope of action for steps towards a sustainable company.

Established economics is almost exclusively occupied with the functioning of market and price mechanisms as economic interaction system. Matters concerning the economic order are seen as given normative decisions in neoclassical macroeconomics and the related microeconomic business theory (Dyllick 1989, p. 80). Recent advancements of these theories are the approaches of property rights theory and institution economics (see annotations). These do no longer confine their studies to the efficiency of exchange contracts and of personal action but also include the issues of external effects and of an efficient shaping of institutional framework conditions (economic order). Thereby, they abandon the methodological individualism of mainstream economics with its neoclassical profile and overcome the reduction of the idea of institutional efficiency to a concept of exchange contracts among market partners (Ulrich 1988, p. 199).

So New Political Economics supplements the institutional level of the exchange contract by the higher level of the social contract. Determination of regulative rules for the validity of exchange contracts – rules that have also to adequately consider the interests of 'external' persons concerned – is of a systematically higher order than conclusion of such contracts (Buchanan 1977, pp. 11ff). Neoliberal and ordoliberal economic concepts both see the necessity of a politico-economic framework order. But neoliberalism does so for mere economic reasons. It focuses on market efficiency as the total criterion. Optimum framework conditions should ensure and increase market efficiency. Ordoliberalism sees the economic order as only a partial social order, which formulates politico-ethical targets (social justice, environment protection etc.) as an extra-economic framework (Ulrich 1994, pp. 22ff.). While neoliberal and ordoliberal concepts conceive political targets and legal provisions as interaction system, they both completely disregard the control potentials and actual control effects of the public, of public opinion, and of media for market processes and entrepreneurial action. But it is exactly public debate that plays a major role in the definition of goals and indicators for a sustainable development.

The theory of external interaction systems, based on thoughts of New Political Economics and institutional management theory, fills an essential gap here (Dyllick 1989). It distinguishes the three external interaction systems market, politics, and public. It tries to identify the most relevant spheres of influence to which companies are exposed in their action and their ability to self-assertion. The influence exerted on companies by market and social environment and, vice versa, the influence coming from companies, may be described and analyzed by the structural model presented in Fig. 1.

While the above-presented stakeholder concept sees things in an actor-oriented perspective, the theory of external interaction systems adopts a functional view. Often, stakeholders are understood as 'representatives' of market, politics, and public, or equalled to these spheres of influence. Such equalling is misleading, though, since companies as well as stakeholders employ different interaction systems in their interactions. For example, a fiscal authority once acts as government control organ in relation to a furniture company (tax control), exerting its influence through the interaction system politics. And at another time, this same authority buys office furniture and exerts its influence on the furniture company via the interaction system market. This fact is particularly obvious with environmental associations like Greenpeace, which in some situations try to exert influence on companies by actions with great publicity, like in the case of the oil production platform Brent Spar (public), at another time by objecting to licence procedures (politics), or sometimes even by active product development, like in the case of a CFC-free refrigerator (market).

The structural model to describe and analyze relations between companies and stakeholders takes into account that external interaction systems cannot be explained by describing only a unilateral influence on the part of market, politics, or public on companies. Companies in their turn also make efforts to influence interaction systems and framework conditions according to their priorities, through marketing, lobbying, and public relations. Nevertheless, we will first look at the influences of external interaction systems on companies.

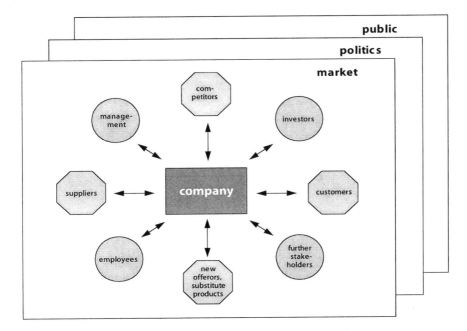

Fig. 1.   Structural model of the relations between companies and stakeholders
Source: the author, following Belz (1995), p. 10

The Interaction System Market

The theory of external interaction systems sees 'market' as an exchange system regulated by the control mechanism of prices and competition. Customers can influence companies and their offers by demands, price negotiations, quality requirements, delivery conditions, or the purchase of alternative products from competitors. In terms of contract theory, market relations are based on "binding explicit or implicit agreements about the exchange of goods and services among persons who consent to these agreements since they see them as improving their situation" (Wolff 1995, p. 38). In this sense, markets are understood as networks of short-term contracts between economically and juristically autonomous economic units. Commercial private interest is acknowledged as legitimate motivation for action, and money as its chief action medium.

There is a great variety of possibilities to influence sustainability of companies by market relations. Among these are, for instance, consumers' decisions in favour of energy-saving and long-lived household appliances, preference for buying alternative food produced by organic farming, eco-ethical investment funds, and also initiatives of big trade companies to ensure the ecological quality of their purchased products. Even authorities can influence companies' compliance with social and ecological requirements by means of their allocation and buying policies.

A precondition for employing the market mechanism to improve a company's environmental performance is transparency of the market. Transparency is reached by a standardization of performance criteria, data collection, data evaluation, and reporting practice. Respective starting points are provided by efforts to standardize the evaluation of environmental performance (see Chapter A.6), by activities related to benchmarking (see Chapter B.8), as well as by attempts to warrant meaningfulness and reliability of companies' environmental reporting (see Chapters A.5, B.6, B.7).

Furthermore, the interaction system market can be used by companies to ecologically distinguish their competitive profile. In that way, the market mechanism may contribute to an ecological structural change. But the correlation of ecology and competitiveness in its ambivalence (cost problem versus distinction opportunity) needs to be profoundly analyzed and purposefully shaped (see Chapter A.3). The credit and investment market represents a further crucial starting point for promoting a 'green reorientation' (see Chapter A.7).

The Interaction System Politics

The interaction system politics has to be understood as system of governmental authorities with democratically legitimized internal decision-making. The representatives elected by the people, the sovereign, constitute communal, regional, state, federal, and transnational parliaments. Apart from legislative power, also executive and judicative power are vested with sovereign rights. In this sense, the interaction system politics hence also comprises the legal field as probably most important form of expressing political control. Its legitimate action motivation consists in general welfare and 'public interest'. Action medium is the political influence.

The relation between companies and executive power may be characterized as a 'sovereign-determined compulsory exchange'. Exchanged are government services (interior and exterior security, traffic infrastructure etc.) against levies to be paid by companies and their compliance with legal regulations. Practice shows that companies and economic associations considerably influence legislation by lobbying. So, the interaction system politics includes 'power arenas' in which processes of interest-guided negotiations take place.

With respect to sustainable development, the major task of the interaction system politics consists in establishing – by means of laws, compliance control, and jurisdiction – market-economic framework conditions that urge companies to do sustainable business. In view of the preservation of the natural basis of our lives, existing environmental regulations have to be supplemented by further international and national environmental goals, environmental quality goals, and environmental action goals (see Chapter A.4). As part of a national sustainability strategy, the inquiry commission of the German Bundestag "Protection of Humans and the Environment" has, for instance, produced a comprehensive catalogue of environmental quality goals and environmental action goals for the problem field of 'soils' (Enquête 1997).

Moreover, the inquiry commission has elaborated four basic rules which should ensure that later societies are not worse off with respect to environmental quality and provision of natural resources than we are today. The German Expert Council for Environmental Issues SRU has complemented them by a fifth rule (SRU 1994, p. 41). In the years to come, these rules will serve as an orientation for the interaction system politics and thus for further legal regulations:

1. The rate of exploiting renewable resources must not surpass the rate of their regeneration. This corresponds with the demand to preserve the functionally defined ecological real capital.
2. Non-renewable resources are to be used only to that extent to which a physically and functionally equivalent substitute in form of renewable resources or in form of higher productivity of renewable and non-renewable resources is created.
3. Immissions into the environment are to orient by the load-carrying capacity of the environmental media. This means that all functions have to be taken into account, among them not least the more sensitive regulation mechanisms.
4. There has to be an adequate correlation between the tempo of anthropogenic immissions into the environment or interference with it and the tempo of those natural processes that are relevant for the response capacity of the environment.
5. Dangers and unjustifiable risks for human health through anthropogenic impacts have to be avoided. This brings up the aspect of risk prevention, which is of importance in the first place for risk technologies like nuclear technology, synthetic chemistry, and genetic engineering (Rochlitz/von Gleich 1997, p. 27).

It is a major task of the interaction system politics to concretize the idea of sustainable development (BMU 1996, p. 9), to integrate requirements into individual policy fields, and to elaborate and pass governmental environment plans (Enquête 1997). While in Germany such a green plan is still lacking, the Netherlands have already produced one. This plan represents an important orientation framework for the efforts of industrial sectors and companies to protect the environment (see Chapter A.4).

Apart from the 'classical' instruments of environment policy (rules and interdictions on the one hand, economic instruments like levies, taxes, and charges on the other hand) a new generation of environment-political control instruments has come up in recent years, namely the so-called indirect regulations. For one thing, these instruments are supposed to reduce the extent of governmental control tasks, and in addition they are to define the scope of economic activities of industry and trade in a way that creates incentives for a dynamic improvement of processes and products. Examples of indirect regulations are official eco-labels for products (Rubik/Teichert 1997), or the Environmental Management and Audit Scheme EMAS of the European Communities (Fichter 1995). Particular importance for the promotion of sustainable development will have to be assigned to information regulations. Among those we have to count, for instance, the US Toxics Release Inventory (TRI) (see Chapter A.5).

The Interaction System Public

In everyday language, 'public' is used as a synonym for certain actors, but that usage is not suitable for an analysis of the interdependence of companies and 'public' (Zerfaß 1996, p. 196), since any actor can draw on publicity to influence other actors. More fruitful in this respect is an understanding of 'the public' as a communication arena. In this arena, which is more or less accessible to everybody, actors turn to a public or at least are exposed to public observation. Apart from a multitude of partial publics, the socio-political public dominated by the mass media plays a major role. The interrelation of speakers, media, and public in this context can be characterized as an exchange relation. Speakers expect publicity for their representation of issues and opinions, the media expect issues and opinions by means of which they can gain themselves public attention and consent.

The public exerts its most important control effects on companies by making company-relevant information available and publicly known, by considering a company's activities under the perspective of public welfare, and by forming a public opinion concerning the company. A pressure for change or legitimization arises whenever the disclosing of company-relevant information makes clear that a company's action and the ensuing effects do not match with the demands of influential social stakeholders.

This may be the case with illegal practices (neighbours, executive/enforcement authorities, jurisdiction), with below-average profits (shareholders, investors), with politically controversial issues like nuclear energy (political parties, environmental associations), or with ethically motivated demands of individual stakeholder groups (animal protectors, human rights organizations). A disapproving public opinion increases public pressure. It is somehow the judgment of an 'invisible parliament' and may retroactively affect in various ways opinion formation of the population, of social stakeholders, and of a company's employees.

Conflicts and confrontations between companies and external stakeholders usually do not emerge within some days, they go through a certain 'life cycle'. According to the stage of the life cycle, public and media play a different role. As a rule, there are five stages in the career of a public issue (Dyllick 1989, pp. 231ff.):

1. latency phase (first events occur, experts show interest, there are latent differences between events and expectations),
2. emergence phase (events occur more frequently, specialized media report about them, interest groups form, social requests take shape),
3. upswing phase (politicalization, mass media report with increasing frequency, there are the first politicians to take up the subject),
4. maturity phase (political parties give their opinion, a regulation is politically prepared, mass media look for other subjects),
5. downswing phase (the new regulation is enforced, behaviour is controlled, non-compliance is sanctioned).

In public mediation of sustainability demands, individual actors play different roles. While significant influence is exerted by critical citizens, scientists, and specialized journalists during the latency and emergence phases, mass media play a crucial role during upswing phase and politicalization of social requests. They have, so to speak, the function of the 'illumination system' of the public arena. Of prior importance during the maturity phase are in most cases governmental political actors. And during the downswing phase, the important actors are control bodies like enforcement authorities, or else environmental associations which take a controlling look at the implementation of a new regulation.

Obviously, the public and the media may have an important role in promoting and enforcing requirements of sustainability. A necessary condition for this is, however, a greater transparency about social and ecological burdens weighing on companies, and about the companies' respective performance. Environmental reporting is of great relevance here. Apart from producing obligatory reports (emission reports, waste water values, waste balances etc.), a growing number of companies today employ voluntary environmental reporting as a means to improve their image and to gain profile in the market. It is not surprising that they want to present themselves from their best side and so primarily publish data and information that show their environmental performance in a favourable light. Accordingly, sufficient transparency has not yet been realized. Clear legal regulations for a company-related environmental reporting (see Chapter A.5, B.7) as well as standardized data collection and evaluation and standards for publication are therefore necessary (see Chapter B.6).

In addition, rating-agencies will be assigned an important future publicity task since they evaluate and publish companies' environmental and social performance (Fichter/Grünewald 1995). Moreover, serious changes will have to take place in mass media. Their present fixation on sensations (malfunctions etc.) and catastrophes (oil tanker accidents etc.) will have to give way to a new structure and culture of issues and ways of reporting (Rolke et al. 1994, Angres 1996). Table 1 gives a survey of the characteristics of external interaction systems.

Table 1. Characteristics and control mechanisms of external interaction systems

	external interaction systems		
	market	politics	public
characteristics	exchange system: for goods and services, with individual utility maximization as legitimate basis of action	authority system: democratically legitimized, vested with hierarchic sovereign power	communication arena: basically accessible to everybody, issues and opinions are exchanged against attention and consent
control mechanism	• price • demand • competition	• legal regulations • compliance control • jurisdiction	• publicity • examination of social claims • public opinion
influence of companies on interaction system by means of ...	price, quality, market communication, contracts, cooperation with other companies etc.	lobbying, activities of associations, political and juridical comments etc.	public relations, press releases, dialogues with opinion leaders, company publicity etc.

Use and Extension of the Scope for Entrepreneurial Action

It has been shown in the preceding observations that with their activities companies are subject to the influence of external interaction systems and that, in turn, companies try to influence and use market, politics, and public according to their goals and interests. This means, from an entrepreneurial view, that companies have to achieve a double task: on the one hand exploit the given scope for action according to own goals, and on the other hand extend that scope. Obviously, a company's potential economic, ecological, and social goals do not simply form a harmonious unity, they are of a sometimes considerably conflicting nature. From the view-point of sustainable development, a company's task is to increase the range where goals harmonize with each other and to influence framework conditions in that sense.

If we hint here at possibilities for extending the field where business, ecological, and social goals meet, this is but an abstract orientation which needs further concretization. A first step in this direction would be to determine action principles on which a sustainable company has to be based. On the one hand, such principles would have to reflect the social role of companies as multifunctional units creating additional value, on the other hand, they would have to consider what loads can maximally be imposed on ecological and social systems. In addition, procedural and regulative requirements would have to be taken into account that are preconditions for a company to 'communicate' and 'hold its own' in its market and social environment. Considering this, we will propose the following principles:

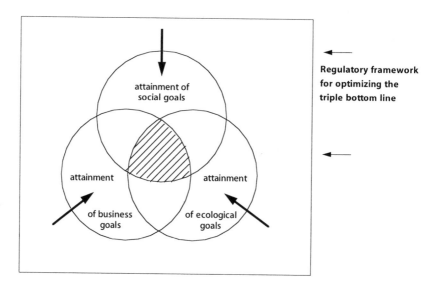

Fig. 2. Model of intersecting sets representing the interrelation of business, ecological, and social goal attainment in companies
Source: the author, following Ulrich (1991), p. 10

Seven Principles for the Sustainable Company

Principle of Performance: A company's essential social contribution are its products and services. If these show optimum performance as regards their orientation by needs, their utility function, their safety, their economic efficiency, and their environment-friendliness, this will mean a significant contribution to sustainability. A sustainable company does not confine its performance and innovations to increasing the eco-efficiency of existing products and processes ('efficiency revolution'). Instead, it orients them by the question which social needs and fields of needs can best be met with which utility functions. Innovation strategies orient in this way. Company performance is to be optimized and evaluated with respect to the whole chain of value added or the whole life cycle of a product or a service.

Principle of Precaution: Dangers and unreasonable risks for human health and natural bases of life due to anthropogenic interference have to be avoided. The sustainable company takes into account that environmental impacts of numerous substances or technologies are not or only insufficiently known and may hence lead to unintended consequences of action. In spite of the international struggle for market shares, a sustainable company orients by the principle of precaution with respect to research, development, construction, and application of products, substances, and technologies. It also supports efforts to achieve internationally standardized and binding legal provisions for risk prevention. Not technical feasibility but social responsibility determines its action.

Principle of Prevention: The sustainable company helps as best as it can to guarantee that later generations are not worse off than we are today with regard to environmental quality and supply with natural resources. To do so, it avoids both exploitation of resources that goes beyond legally determined upper limits and use that obviously exceeds the depletion rate of renewable resources. It avoids consumption of non-renewable resources or at least restricts the respective volume according to the rate at which a physically and functionally equivalent substitute in form of renewable resources or higher productivity is created. It orients its immissions into the environment by the load-carrying capacity of the environmental media and their partly very sensitive regulation mechanisms.

Principle of Dialogue: Sustainable development is a social process of search and achieving an understanding. It consists of balancing different social interests and perceptions. Conflicts of values and interests are an integral part of such a process. The communication between a company and its stakeholders has to be dialogue-oriented so as to build up lasting potentials for understanding. Dialogue orientation, in contrast to the persuasive communication style of market communication and traditional public relations concepts (manipulate, persuade), is characterized by an argumentative communication style (convince). Understanding-oriented action is a major regulative of a sustainable development.

Principle of Development: With a sustainable development, not only market demands on the company change but also politico-juridical and social requirements. In this context, sustainability of a company is to be understood not as a state but as a dynamic process of continuously identifying a new ecological, social, and economic chances and risks. If a company wants to survive in this process of change, it has to be able to develop and learn. The important element of organizational learning is learning to solve problems. This 'learning how to learn' is a suitable means to ensure the ability for innovation and change. Strategic, structural, cultural, and personal development are major starting points for organizational learning. The ability to change may also presuppose 'dissolving company borders' as well as developing cooperative solutions (strategic alliances, joint ventures, etc.) and company networks.

Principle of Compliance: Market development and profit interests with their proper dynamics need clear regulative framework conditions. It goes without saying that a sustainable company complies with legal provisions. In addition, it orients by eco-politically and socio-politically set priorities and targets as are laid down, for instance, in international contracts, declarations, or protocols and as are formulated as national, regional, or local goals for environmental quality and environmental action.

Principle of Responsibility: Sustainable development does not only presuppose increased eco-efficiency and consistency (adjusting products, production, and services to natural cycles) but also an altered orientation with regard to prosperity and quality of life. With their products and their publicity, companies shape models and life styles, hence represent a normative social instance. The sustainable company critically reflects customers 'models and consumers' life styles in the context of its normative and strategic management, and contributes as best as it can to confinement and moderation (sufficiency).

Steps Towards the Sustainable Company

Sustainability principles have to be translated into reality on several levels, on the regulative level (framework conditions for the market economy) as well as on the level of company policy (normative management), on the strategic level (strategic management) as well as on the operative one (operative management). From an entrepreneurial perspective, taking into consideration practical experiences and insights of research on strategic management, seven starting points can be listed that are of overall and essential importance for a company's efficiency and competitiveness. With the help of these starting points, principles and goals of sustainability can be operationalized and translated into concrete steps towards the sustainable company.

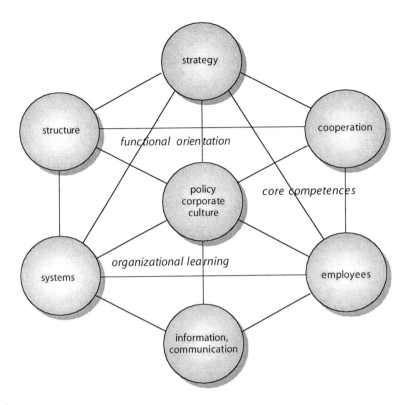

Fig. 3. Strategic starting points for a sustainable management
  Source: the author, following Waterman, Peters and Phillips (1980), p. 17

In practice, these seven starting points for a sustainable management are closely interlinked and represent an entire entrepreneurial system. Nevertheless, they can be discussed individually, and autonomous implementation steps can be formulated. Interdependence with other elements has to be considered in implementation, though. We will now shortly present the starting points for a sustainable management and give examples of individual implementation steps or implementation instruments.

Policy, Corporate Culture

Sustainable management presupposes a company's will and intention to substantially contribute to sustainable development. This means that the company has to reflect on principles and goals of sustainability and root them in its basic normative orientation. This is not merely achieved by the company management determining company guidelines or principles. Essential is the integration of sustainability goals into a company's actually lived system of values and standards (corporate culture).

Participative approaches to elaborate company policies and guidelines may be of major help here (Niggemeyer 1993). If employees at different hierarchic levels are included in this process, current opinions and values of employees and managers can be known, and their ideas can be included into the formulation of visions, basic principles and goals (Lehmann/Hüther 1995).

Principles of sustainability must not remain abstract, instead, they have to become 'tangible' for employees and managers in their everyday practice. For this purpose, principles have to be concretized by strategic and operative management. Only by transforming basic principles into definite environmental programmes and department-related or person-related targets, e.g. by means of company plans or individual agreements on targets, can sustainability become tangible and realizable. The value 'sustainability' is less a matter of words than a matter of deeds. This is of particular relevance for the company management, which has to have itself measured by this demand.

Strategy

Strategic management is of crucial importance with respect to preventive environment protection. Numerous economic, social, and ecological problems could be avoided if chances and risks were included into consideration and evaluated at an early stage of research and development or of portfolio management. Strategic management theory offers a wide range of instruments to be used for sustainable company strategies (Hinterhuber 1996). Of decisive importance for developing environment-friendly company strategies is the explicit and conscious consideration of criteria and models of sustainability as well as orientation by social fields of needs, like nutrition, health, construction and housing, or mobility. The fact that material goods are today understood as the absolute and only means to satisfy needs has, in recent decades, induced a considerable exploitation of the environment and caused correspondent ecological problems. Therefore, company strategies have to focus on the question how social needs can effectively and efficiently be satisfied. Such a perspective emphasizes functions, not (material) goods, as of priority for new business fields. So, mobility demands need not necessarily be satisfied by means of an own car, but may draw on function-oriented services (car rental, car sharing etc.). Functional orientation opens up new market opportunities and helps to dematerialize the satisfaction of needs (Pfriem 1995).

Structures

The rapid change of conditions of global competition has caused an increasing dissolution of company-internal hierarchies in recent years, and hence a 'modularization of companies'. "Modularization means restructuring the company organization on the basis of integrated, customer-oriented processes into small units (modules) easy to be surveyed. These are characterized by decentralized decision competence and result-oriented responsibility." (Picot/Reichwald/ Wigand 1996, p. 201). Modularization is advanced by the breakneck development of new information and communication technologies and the increased tendency towards the buyer's market and thus towards customer-oriented and individual manufacturing.

Modularization trends can be made useful for a more sustainable way of doing business. In social respect, trends towards order-oriented production allow an increased application of concepts like job rotation, job enlargement, job enrichment or the establishing of (partly) autonomous groups of employees. A larger scope for action and enlargened decision competence will lead to a so-called 'empowerment' of employees. These tendencies will support the employees' self-development and increase their performance motivation.

Decentralized company structures involve the risk of 'organized irresponsibility'. Hence, on the one hand, environmental requirements have to be purposefully integrated into central control units (holdings, combines, central areas), and on the other hand, they have to be rooted in the largely autonomous company areas. Apart from the usual functional-additive structures of environment protection (environmental departments and the staff position of an environmental officer etc.), sustainability requirements have to be integrated into business processes and made also the task of line managers, up to rooting special environmental tasks in job descriptions and individual agreements on targets. The crucial task here is to organize a preventive environment protection (Antes 1996).

Systems

Suitable for structuring and implementing complex cross-section-oriented company tasks are management systems. Since the 80s, management systems in the form of quality interaction systems, environmental management systems, and systems concerning occupational safety and security of facilities have increasedly been applied in company practice. Most recently the international standard "Social Accountability 8000" has been developed. For e.g. the largest mail order company of the world, the Otto corporation with its headquarters in Hamburg, Germany, has been starting to implement a social management system according to this standard. This systems will be integrated in Otto's supply chain management.

With the help of management systems, cross-section-oriented targets and tasks can systematically be integrated into operational processes, and their realization can be ensured. The essential function of management systems is to organize a process of continuous improvement.

Particularly appropriate to ensure a dynamic process of ecological improvement is the concept of eco-controlling. As a management-supporting system, it covers functions regarding analysis, planning, coordination, control, and communication in the context of environmental management (Fichter 1995a, p. 66). On the basis of regular company eco-balances, potentials for improvement can be identified and environmental goals and measures can be planned and realized (BMU/UBA 1995; Gallert/Clausen 1996; Hallay/Pfriem 1992).

Information and Communication

The importance of information and communication in economic life and their relevance for a company's competitiveness has continuously increased in the recent decades. New requirements for a company's information and communication management have arisen due to globalization of the markets for goods, labour, and information, due to decentralization of companies, due to lower vertical range of manufacture, and due to the breakneck spreading of new information and communication technologies (Picot/Reichwald/Wigand 1996).

In view of lower vertical range of manufacture and decentralized production structures, quality and ecological data have increasedly to be exchanged among companies to meet quality and environment protection requirements. This holds both for manufacturing-related and material data and for information about quality and environmental management systems. Therefore, reporting between suppliers and customers and the exchange of information along chains of value added have to be explicitly considered in the context of a sustainable management. Environmental reporting systems and material flow management represent important starting points here (Clausen/Fichter 1994, see also Chapter B.10). Precondition for an inter-company information management and the optimization of a company's environmental performance is that kind of information that can be gained with the help of instruments like environmental cost accounting (Fichter/Loew/Seidel 1997, see also Chapter B.9), environmental performance evaluation (see Chapters A.6, B.8), environmental performance indicators (Loew/Hjálmarsdóttir 1996), and life cycle assessment (see Chapter B.3).

During the recent decades, risky side-effects and consequences of industrial activities as well as risk sensitivity of the population have also increased. This has resulted in an increased 'public exposure' of companies (Dyllick 1989). Reach and interference level of industrial activities conflict more and more with social requests. So, company communication has to orient more intensively towards dialogue and use an argumentative style in discussions with citizens' initiatives, environmental organizations, and consumer associations. Examples for entrepreneurial action that orients towards mutual understanding are discussion groups with neighbours, environment forums, and environment counsellors (Fichter/Loew 1997, p. 106).

Employees

Organizational learning is based on individual processes of learning. A great variety of concepts and instruments have been developed since the 80s to foster professional eco-knowledge and qualification with respect to environment protection (Nitschke et al. 1996; Hopfenbeck/Willig 1995). Professional education and training, the integration of ecological subjects into university curricula, and the ability for transdisciplinary cooperation play a major role if we mean to lastingly promote employees' environmental awareness and environment-friendly action.

Moreover, participation and qualification of employees in the context of environmental management systems are of prior importance (Fichter 1995b). This means that a continuous ascertaining of the need for environmental training and qualification and the implementation of corresponding measures have definitely to be rooted in a company's management system (Petersen 1997). Ecological action does, however, not only presuppose the respective awareness and qualification of individuals. In addition, systems of incentives and evaluation (suggestion scheme, bonus scheme, employee rating) as well as working and participation structures (scope of action at the work place, team concepts, etc.) have to be designed in an appropriate corresponding way.

Cooperation

Independent of sustainable development, it can generally be noticed in business practice that traditional company structures and company borders dissolve and change towards 'symbiotic connections' with external partners. Meanwhile, there is a multitude of different forms of symbiotic partnerships. Among them are, for instance, strategic alliances, joint ventures, and company networks (Aulinger 1996; Picot/Reichwald/Wigand 1996; Sydow 1992). Yet symbiotic connections do not only exist among companies, for a couple of years already they have also occurred among companies and environmental associations or institutions. Examples are the cooperation of the 'Bund für Umwelt und Naturschutz' (union for environment and environment protection BUND) with firms like the German department stores group Hertie or the German Federal Railways (Baßfeld 1997), and the cooperation of the German Eco-Institute with the chemical combine Hoechst (see Chapter B.11). In the context of sustainability, such cooperation of different actors represents considerable opportunities (Ökologisches Wirtschaften 2/1997). Different forms of cooperation may be used, among them the following:

- market-oriented vertical company cooperation along chains of value added
- market-oriented horizontal company cooperation within one industrial sector
- politically oriented cooperation between companies or their associations and governmental actors
- publicly oriented cooperation between companies and organizations like environmental or consumer associations
- lateral cooperation as a mix of above-listed forms of cooperation (Schneidewind/Hummel/Belz 1997, p. 42).

Conclusion

The great number of starting points, concepts, and instruments that can be used for steps towards a sustainable company reflects the diversity and complexity of the task. The path to sustainable development is neither trodden in one big leap nor is its definite end ever reached. This way is a dynamic process full of chances and risks. It is important to make actual steps and at the same time not to lose sight of the overall idea.

Annotations

The property rights theory is essentially concerned with action and disposition rights regarding goods. These rights define and hence also confine what rights an individual has in dealing with a good. While neoclassical microeconomic theory is based on the idea of profit maximization as a guiding maxim for entrepreneurial action, property rights theory concedes economic actors a more open range of goals. Utility maximization is not restricted to achieving material benefits. For instance, the individual can maximize his or her individual utility as an altruistic benefactor. Hence the concept of individual utility maximization is in itself free of any specific positive or negative meaning.

New institution economics stresses information and communication as important factors for coordinating economic activities. It focusses on 'institutions' that serve the rationalization of information and communication processes. In doing so, it studies institutional outcomes like contracts, organizational structures, language, money etc. for their effects on human behaviour as well as for possibilities of an efficient design (Richter 1994).

References

Angres, V. (1996): Schritte zu einer nachhaltigen, umweltgerechten Entwicklung – Überlegungen und Beiträge aus dem Bereich der Medien, in: BMU – Bundesministerium für Umwelt, Naturschutz und Reaktorsicherheit (ed): proceedings of the conference "Schritte zu einer nachhaltigen, umweltgerechten Entwicklung", Bonn, July 1, 1996.

Antes, R. (1996): Präventiver Umweltschutz und seine Organisation in Unternehmen, Wiesbaden.

Aulinger, A. (1996): (Ko-)Operation Ökologie, Kooperationen im Rahmen ökologischer Unternehmenspolitik, Marburg.

Baßfeld, J. (1997): Aktive Mitarbeit an der Umgestaltung der Wirtschaft, Wirtschaftskooperationen beim Bund für Umwelt und Naturschutz (BUND), in: Ökologisches Wirtschaften 2/1997, pp. 25ff.

Belz, F. (1995): Ökologie und Wettbewerbsfähigkeit in der Schweizer Lebensmittelbranche, Bern, Stuttgart, Vienna.

BMU – Bundesministerium für Umwelt, Naturschutz und Reaktorsicherheit (ed) (1996): Umweltpolitik – environmental expert opinion 1996 of the German Expert Council for Environmental Issues, short version, Bonn.

BMU/UBA – Bundesministerium für Umwelt, Naturschutz und Reaktorsicherheit und Umwelbundesamt (ed) (1995): Handbuch Umweltcontrolling, Munich.

Buchanan, J.M. (1975): The Limits of Liberty, Between Anarchy and Leviathan, Chicago, London.

Clausen, J.; Fichter, K. (1994): Wissenschaftlicher Endbericht zum Projekt Umweltberichterstattung, Osnabrück, Berlin.

De Man, R.; Ankele, K.; Claus, F.; Fichter, K.; Völkle, E. (1997): Aufgaben des betrieblichen und betriebsübergreifenden Stoffstrommanagements, text volume of the German Federal Environmental Protection Agency UBA, Berlin.

Dyllick, T. (1989): Management der Umweltbeziehungen, Öffentliche Auseinandersetzungen als Herausforderung, Wiesbaden.

Enquête-Kommission "Schutz des Menschen und der Umwelt" of the 13th German Bundestag (1997): Konzept Nachhaltigkeit, Fundamente für die Gesellschaft von morgen, Bonn.

Fichter, K. (1998): Umweltkommunikation und Wettbewerbsfähigkeit, Wettbewerbstheorien im Lichte empirischer Ergebnis zur Umweltberichterstattung von Unternehmen, Marburg.

Fichter, K. (ed.) (1995): Die EG-Öko-Audit-Verordnung, Mit Öko-Controlling zum zertfizierten Umweltmanagementsystem, Munich, Vienna.

Fichter, K. (1995a): Der Ablauf des Gemeinschaftssystems: mit Öko-Controlling zum zertifizierten Umweltmanagementsystem, in: Fichter, K. (ed.) (1995): Die EG-Öko-Audit-Verordnung, Mit Öko-Controlling zum zertfizierten Umweltmanagementsystem, Munich, Vienna, pp. 55–70.

Fichter, K. (1995b): Ermittlung des Informations- und Bildungsbedarfs im Umweltschutz, in: Fichter, K. (ed.) (1995): Die EG-Öko-Audit-Verordnung, Mit Öko-Controlling zum zertfizierten Umweltmanagementsystem, Munich, pp. 109–118.

Fichter, K.; Grünewald, M. (1995): Öko-Rating, Ansätze zur ökologischen Unternehmensbewertung, IÖW discussion paper 32/95, Berlin.

Fichter, K.; Loew, T. (1997): Wettbewerbsvorteile durch Umweltberichterstattung, Ergebnisse einer empirischen Untersuchung über Stand, Aufwand und Nutzen der Umweltberichterstattung von Unternehmen in Deutschland, Berlin.

Fichter, K.; Loew, T.; Seidel, E. (1997): Betriebliche Umweltkostenrechnung, Methoden und praxisgerechte Weiterentwicklung, Berlin, Heidelberg, New York.

Freeman, R.E. (1983): Strategic Management – The Stakeholder Approach, in: Advances in Strategic Management, vol. 1, 1983, pp. 31–60.

Gallert, H.; Clausen, J. (1996): Leitfaden Öko-Controlling, Düsseldorf.

Hallay, H.; Pfriem, R. (1992): Öko-Controlling, Frankfurt, New York.

Hinterhuber, H.H. (1996): Strategische Unternehmensführung, I. Strategisches Denken, Vision, Unternehmenspolitik, Strategie, 6th reviewed and complemented edition, Berlin, New York.

Hopfenbeck, W.; Willig, M. (1995): Umweltorientiertes Personalmanagement, Landsberg/Lech.

Lehmann, S.; Hüther, S. (1995): Umweltpolitik und Umweltleitlinien, in: Fichter, K. (ed.): EG-Öko-Audit-Verordnung, Mit Öko-Controlling zum zertifizierten Umweltmanagementsystem, IÖW-Schriftenreihe 81/95, Berlin, pp. 17–28.

Loew, T.; Hjálmarsdóttir, H. (1996): Umweltkennzahlen für das betriebliche Umweltmanagement, IÖW-Schriftenreihe 99/96, Berlin.

Niggemeyer, G. (1993): Die Entwicklung und Umsetzung umweltorientierter Unternehmensleitlinien, IÖW-Schriftenreihe 68/93, Berlin.

Nitsche, C.; Fichter, K.; Loew, T.; Scheinert, K.; Schöne, H. (1995): Berufliche Umweltbildung – Wo steckst Du?, Ergebnisse einer Untersuchung in 28 Institutionen, Bielefeld.

Ökologisches Wirtschaften (2/1997): special issue "Kooperation im Umweltschutz", edited by the Ecological Economics Research Institute and the Ecological Economics Resarch Union (IÖW/VÖW), Berlin, March/April 1997.

Petersen, S. (1997): Organisation von Mitarbeiterschulung und -information im Rahmen eines Umweltmanagementsystems, IÖW-Schriftenreihe 120/97, Berlin.

Pfriem, R. (1995): Unternehmenspolitik in sozialökologischen Perspektiven, Marburg.

Picot, A.; Reichwald, R.; Wigand, R. (1996): Die grenzenlose Unternehmung, Information, Organisation und Management, 2nd updated edition, Wiesbaden.

Richter, R. (1994): Institutionen ökonomisch analysiert, Tübingen.

Rochlitz, J.; Gleich von, A. (1997): special expert opinion of Prof. Dr. Jürgen Rochlitz and Prof. Dr. Arnim von Gleich, in: Enquête-Kommission "Schutz des Menschen und der Umwelt" of the 13th German Bundestag (1997): Konzept Nachhaltigkeit, Fundamente für die Gesellschaft von morgen, Bonn, p. 27.

Rolke, L; Rosema, B.; Avenarius, H. (eds.) (1994): Unternehmen in der ökologischen Diskussion, Umweltkommunikation auf dem Prüfstand, Opladen.

Rubik, F; Teichert, V. (1997): Ökologische Produktpolitik, Stuttgart.

SRU – Der Rat von Sachverständigen für Umweltfragen (1994): Umweltgutachten 1994, document of the German Bundestag 12/6995, Bonn.

Sydow, J. (1992): Strategische Netzwerke, Evolution und Organisation, Wiesbaden.

Treuenfels, C.-A. (1996): Schritte zu einer nachhaltigen, umweltgerechten Entwicklung, Überlegungen und Beiträge aus dem Bereich der Umwelt-/Naturschutzverbände, in: BMU – Bundesministerium für Umwelt, Naturschutz und Reaktorsicherheit (ed.) (1996): proceedings of the conference "Schritte zu einer nachhaltigen, umweltgerechten Entwicklung", Bonn, pp. 27–34.

Ulrich, P. (1980): Plädoyer für unternehmungspolitische Vernunft, in: Management-Zeitschrift IO, 49, No. 1, pp. 32ff.

Ulrich, P. (1988): Betriebswirtschaftslehre als praktische Sozialökonomie, Programmatische Überlegungen, in: Wunderer, R. (ed.): Betriebswirtschaftslehre als Management- und Führungslehre, 2nd completed edition, Stuttgart, pp. 191–215.

Ulrich, P. (1991): Ökologische Unternehmenspolitik im Spannungsfeld von Ethik und Erfolg, Fünf Fragen und 15 Argumente, Beiträge und Berichte No. 47 of the Institute for Economic Ethics, university of St.Gallen.

Ulrich, P.; Fluri, E. (1995): Management, Eine konzentrierte Einführung, 7th reviewed edition, Bern, Stuttgart, Vienna.

Waterman, R.H.; Peters, Th.J.; Phillips, J.R. (1980): Structure is not organization, in: The McKinsey Quarterly, Summer 1980, pp. 2–20.

Wicke, L. et al. (1989): Umweltökonomie, Eine praxisorientierte Einführung, 2nd completed and updated edition, Munich.

Wolff, B. (1995): Organisation durch Verträge, Wiesbaden.

2 Goals for the Sustainable Company

Jens Clausen, Maite Mathes

"Your father, Andrees – don't get me wrong – was too soft a soul. Neglected his purse. A Christian he was, wasn't he? Well .... I wanted to say, Christians we all are – but he, I mean: he really wanted to live as a Christian. He didn't mean to be a Christian just for himself, in his house, with an honourable life, saying grace before the meals and all that, which is nothing to object to. Instead, he wanted to treat workers in a Christian way, buy in a Christian way, sell in a Christian way, in short, do a Christian business."

Gustav Frenssen (1903)

Sustainable company management is increasingly becoming fashionable in business circles. As a fashionable but somehow empty phrase, it is often equalled to ecological company management or even to implementation of the Environmental Management and Audit Scheme. Obviously, it is easier to confine the concept to individual problem fields or actions than to interpret it as a whole and translate it into reality. This is not a new dilemma. As the text quoted above from a novel at the turn of the century shows, even in those times "saying grace and all that" – which would perhaps be "separation of waste and all that" today – was more common than actual Christian (or today: sustainable) buying and selling or a respective personnel policy.

Yet a truly sustainable company must not only be interested in environmental performance, but also in the minimal ethos of a modern free and democratic society, namely the unconditional claim of all people that also in the socio-economic context their nature as human subjects and their personal basic rights be respected (Ulrich 1993).

Therefore, the following text will formulate potential environmental, social, and economic guidelines and goals for the action of a sustainable company, in analogy to the triple bottom-line of sustainability, and put them up for discussion. The presented guidelines for environmental sustainability are the result of a long-standing discussion process. The guidelines for social and economic sustainability, on the other hand, are to be understood as contributing to a current debate. They are meant to be pragmatic steps in the course of attempts to introduce more ethical reasoning into the dynamics of economic rationalisation (Ulrich 1993).

Guidelines for Environmental Sustainability

All forms of economic activity are ultimately based on natural processes, on the advances of a well-functioning environment. At the same time, economic activities interfere with the environment, they shape "nature".

In the context of developing a non-monetary system to ecologically extend macroeconomic accounting, Fischer-Kowalski et al.(1993) reflected on the actual meaning of the notion of environmental impact. They assigned the multitude of respective concepts to four basic paradigms of environment protection. Each of those paradigms has its own structure of argumentation, its own scientific and political tradition, and its own reference public. Each of them refers to a specific basic concept, and each is able to depict important aspects of what an environmental impact might be. The paradigms are not mutually excluding, hence a certain form of environmental impact may be of relevance in more than one paradigm. But none of them can be reduced to another. Nor can they be merged into one "big" paradigm (Fischer-Kowalski et al. 1993).

The guidelines presented below are based on these paradigms. All four guidelines together allow an approximately complete description of what can be meant by "environmental impact of a company".

Guideline 1: "Avoiding Intoxication"

The paradigm "intoxication" is probably the most wide-spread theoretical model. It is derived from the fields of medicine and chemistry. Harmful substances in waste air, waste water, and refuse lead to undesired effects in the environmental media and subsequently in humans and other living beings as well. As a measure to prevent such impacts (which, on principle, but not always correctly, are attributed to too high concentrations of harmful substances), limits and technical rules are put in place to achieve a reduction of toxic immissions.

The determination of limits, this should be noticed, is a process of social standardisation informed by interests and the balance of power. There is no simply scientifically justifiable limit. A political consensus, possibly implicit has to be attained for each individual issue about what risk can and has to be taken by whom. Practical effectiveness of limits and technical directives depends on the organisation of their enforcement, on economic and political arrangements. The scientific debate about (the level of) limits concerning certain substances often disregards these practically relevant key issues (Fischer-Kowalski 1993).

There are two essential difficulties with this theoretical model of avoiding intoxication. Firstly, toxicological technical know-how is systematically developed slower than new substances. Secondly, toxicology seems unable to comprehend systemic links in eco-systems. While one chemist may be able to synthesise 20 new substances in a month, a period of years is presumably needed by one hundred biologists, chemists, and ecologists to identify just the major environmental impacts of those 20 substances. Hence our knowledge about the effects of new substances in the environment necessarily lags behind our ability to produce and use them (von Gleich/Hellenbrandt 1994). A risk-free, responsible assessment of new substances seems possible only with respect to very restricted impact patterns. A comprehensive knowledge of eco-toxicological interrelations is impossible.

Particular attention should be paid to the speed of developing new substances. This speed should not grow beyond certain maximum values which depend on our capacity to assess the connected effects. Like in many other contexts, speed or "the appropriate tempo" is the confining factor over which we should take precautions.

Moreover, even correct observance of limits is no suitable means for avoiding environmental damage in the long run. The quantity of emissions in total gives rise to pollution. Hence observation of limits should be accompanied by limitation of total amounts. "Classical" environment protection with its containment of contamination through restoring technologies becomes more debatable in the context of the discussion about material flows, where emission permits for a limitation of total amounts are also considered.

Hence the guideline of "avoiding intoxication" is an important element of a company policy.

What can a company do? Examples of tangible goals and measures:

- ascertain and evaluate human-toxic and eco-toxic substances in the company
- avoid and reduce the use of human-toxic and eco-toxic substances as raw materials in production and their output as products
- avoid and reduce emissions of human-toxic and eco-toxic substances

---

Guideline 2: "Conserving the Natural Equilibrium"

The theoretical model of a "natural equilibrium" corresponds with the scientific traditions of biology and is also used by climatologists, agroscientists and conservation organizations. In this concept, society operates as an actor interfering deliberately or inadvertently with functional interdependence in natural systems, thereby menacing their capacity of self-preservation. This, in turn, affects society, since certain eco-systems suddenly cease to function in the way they used to (Fischer-Kowalski 1993).

The "natural equilibrium" is often not even a true equilibrium. Once a certain natural equilibrium is destroyed, a new state with new conditions of balance emerges. So "equilibrium" means continuous flowing, is hence a steady state.

The concept of a "natural equilibrium" basically takes into account the systemic character of nature. The equal amount of the same substance may operate as a nutrient or a noxa, according to the circumstances. Two substances, each of which is harmless in itself, may trigger catastrophes once they compound. Among the disturbances of equilibrium states which we have to take into account today are also cultivation of soils with heavy machines, regulation of water courses, import of foreign organisms, or artificial insemination of cows. And clearly, many inconspicuous and sensitive regulation mechanisms of eco-systems are still beyond the scope of scientific description (Fischer-Kowalski 1993).

The difficulty with this equilibrium approach lies in the complexity of balances to be protected. Ultimately, we ignore interrelation in eco-systems, and anyhow their complexity plainly does not allow us to gain complete knowledge. Only a few simple possibilities to preserve eco-systems are known, such as not depriving those systems of their territories (e.g. by soil sealing in road construction), not polluting them thermally (e.g. by waste heat in rivers), with noxae (e.g. by the use of pesticides), or with nutrients (fertilisers). Apart from these elementary rules, hardly any recommendations for action can be given. So the precaution principle seems a convincing maxim here. It may logically be presumed that the probability of an unintentional interference with the natural equilibrium will decrease through applying near-nature technologies, avoiding interference with the regulative structure of living beings, conserving bio-diversity, and orienting by natural tempi and rhythms (von Gleich/Rubik 1996). So business decisions about technologies and products should also orient by these precaution criteria.

What can a company do? Examples of tangible goals and measures:

- give priority to error-friendly technologies of low-level interference when technological decisions are made
- reduce material flows in a preventative manner
- reduce CO_2-emissions
- protect valuable eco-systems
- restore natural habitats

Guideline 3: "Protecting Resources"

In its first two basic principles, the inquiry commission of the German Bundestag "Protection of Humans and the Environment" requires an economical dealing with both renewable and finite raw materials. The basic idea behind this argument is the demand for sustainability of raw materials. Thus finite raw materials must only be used to the extent that their expected range of exploitation can be increased through improved efficiency of use. Renewable raw materials must only be used to the extent that the eco-system producing them is not impaired. This means that the use of solar energy within the framework of what is technically feasible is not subject to restrictions, whereas agriculturally grown raw materials are available only insofar as they can be reproduced by organic farming.

The decisive eco-political question here is which resources are used up at which rate? Does this rate match with or remain even below the rate of their renewal or extension of their range of exploitation?

The guideline for protection of resources is often already found to be in companies' environmental policies.

What can a company do? Examples of tangible goals and measures:

- promote service concepts and shared use of products
- construct products in a material-saving and recyclable way
- increase the eco-efficiency of water and energy use, avoid wastes and recover materials in production

---

Guideline 4: "Respecting the Fellow World"

The fourth paradigm is the orientation by the "fellow world" with its plants and animals. The theoretical model "fellow world" is based on philosophical and ethical traditions that deny humans as a single species the right to dominate all other species at will. These ideas refer to the philosophical debate about the right of all living beings to life and development.

Should we, for instance, allow domesticated animals to live according to conditions that suit their natural behaviour as far as possible, or may we subject them to any potentially cost-effective but restrictive form of raising? Even if, in our short-sightedness, we completely deny some organisms the right to live (e.g. the small-pox virus), this right to life should be generally accepted as "basic right". It has to be noticed that the fellow world paradigm is still that part of the idea of environment protection which is the least able to attain consensus in public debate. A lot of our everyday decisions are rather based on the maxim "subdue the earth". Nevertheless, a political translation of this paradigm into reality is possible by establishing wildlife preserves, by species-adequate keeping of animals, or by the (at present still restricted) ban on animal experiments. The criticism of genetic modification of life itself also reflects this paradigm.

If humans really understand themselves, like the surrounding fellow world, as part of one nature, their respectful dealing with nature will also include respect for themselves and for other fellow humans, and hence is to be assessed as positive for the human species. The processes taken into view as environmental impacts by the "fellow world" paradigm are again of a completely different type than those regarded by the previously presented theoretical models.

What can a company do? Examples of tangible goals and measures:

- provide their animal raw materials from farms where animals are kept in a species-adequate way
- avoid or shorten animal transports, or carry them out in a species-adequate way
- foster biodiversity by fostering rare species, e.g. with appropriate product development and provision of animal and plant raw materials

---

The Four Guidelines Combined

Fischer-Kowalski et al. (1993) illustrate in an impressive way the different lines of argumentation derived from these four guidelines, taking car traffic as an example:

With the guideline of avoiding intoxication, limits for the emission of harmful substances by cars would be required, since cars contribute with a high share to the overall emission of certain toxic substances (CO, NO_x, etc.). Catalysts would be a good solution, as they reduce the respective emissions by about 80%.

With the guideline of protecting resources, it is important that car traffic causes about half of the ultimate consumption of non-renewable fossil fuels. Cars with a more economical use of gasoline or solar cars would be necessary. Catalysts would be a distinctly negative solution, since their production requires the scarce resource of platinum, and their use increases fuel consumption.

With the guideline of a natural equilibrium, above all, the increase in CO_2 due to car traffic is of relevance, as it contributes by 15% to the greenhouse effect. Additionally important are the impacts on certain biological systems, like forests and the hydrological cycle. Catalysts would be of no help, since they do not at all reduce CO_2-emissions, but cars using less energy might be an improvement.

With the fellow world guideline, car traffic would be identified as the major cause of involuntary and meaningless killing of animals and (sometimes non-motorized) humans. Attention would be drawn to the fact that the traffic network cuts habitats of many species (as well as children's space for experience and play) and confines them to areas that do no longer provide acceptable conditions for life (or not even for survival). This problem is neither solved by solar cars nor by catalysts.

As this example illustrates, eco-political argumentation lines differ widely according to guideline. And they lead to different results.

Confusing as this may be, it is useful at the same time. It opens up a legitimate scope for political discussion and decision processes, which cannot be closed again by presumably scientific imperatives or irrefutable "technical logic".

If a social solution wants to be counted as "environmentally-friendly", it will have to stand up to all four guidelines. If it is suboptimal, respective arguments will have to be weighted according to the balance of ideological and political forces, hence according to a "public conscience".

Guidelines for Social Sustainability

Social movements like trade unionism, feminism, or co-operative movement, have a definitely longer history than that of environmentalists. Hence they know quite well the ups and downs, the being "in" and being "out", the alternate situations of being powerful and powerless, of which environmentalists now, in the "eco-pause", have become painfully aware.

No one will expect the debate on sustainability to raise social guidelines for entrepreneurial action that have not been known for long. And yet it is once again indicated, particularly in these times of a dominating neo-liberal global economic order, to emphasise the social responsibilities of the influential actors.

Of course, it would be possible to formulate an additional guideline "information, participation, and co-operation" here. This aspect is added below, as an action principle, because it is relevant also for the dealing with ecological and economic guidelines.

Guideline 1: "Respecting Civil Rights and Liberties"

Although it is not very common to read the general declaration of human rights, passed by the UNO in 1948, it is an important thing to do regarding the subject of social sustainability. Moreover, this declaration represents (at least theoretically) a globally valid standard, hence its realisation or a respective demand could not be interpreted as a "non-tariff trade barrier". It declares the human basic rights to life, freedom and safety, the right to be protected from torture and slavery, the basic rights to citizenship, freedom of movement, and protection of privacy, as well as the rights as a juristic person. Apart from that, it also lays down the freedom of thought and free speech, and the right to equal treatment (men and women, races, religions etc.).

A similar collection of rights will presumably be found in most national constitutions.

Respect for civil rights and liberties is a basic demand on companies. Certainly, most companies in democratic states will show a minimum of constitutional conduct. The critical point with respecting human rights will probably lie in individual aspects, e.g. in business connections with partners in undemocratic states, or in equal rights. Similar to requiring the "application of globally equal standards" in environment protection, even from suppliers, this demand should also be transferred to civil rights and liberties. Multinational enterprises can at least be attributed a certain influence in this field. The New York and London based Council on Economic Priorities recently raised the Social Accountability Standard 8000 based on International Labour Organisation (ILO)-Conventions and founded an Accreditation Agency (CEPAA 1997). It will such be possible to audit against social standards as well.

What can a company do? Examples of tangible goals and measures:

- integrate the respect for human rights into the "compliance audit" of all sites, e.g. SA 8000
- consider the respect for human rights in its supplier audits, particularly with suppliers from undemocratic states, e.g. SA 8000
- ensure equal rights of men and women, of persons with different religions, nationalities, and races
- in the long term, consider these groups proportionately when appointing persons to management positions (quota)
- employ handicapped persons and take them into consideration as customers
- take into account requirements and potentials of older persons in their integration into professional life

Guideline 2: "Fostering Safety and Health"

Health and safety are also the subject of the general declaration of human rights (articles 3 and 25). As goals of entrepreneurial action they are largely undisputed, but with regard to the way of making necessary respective provisions, considerably different opinions are found in society.

So, companies are responsible for occupational safety, for the safety of people living in the neighbourhood of their sites, and for safety with regard to dangerous transport. Statistics regarding accidents and health provide information about the success of a respective commitment. Existing safety equipment and safety directives, recreation institutions and break regulations indicate some precautionary efforts. Yet it is apparently a matter of debate whether the goal of occupational safety and security of facilities has to be translated into the demand for an error-friendly technology the malfunctions of which cannot cause any irreversible damages, or into the demand for risk control. Sustainable companies should take the demand for error-friendliness seriously and avoid technologies with high-level interference into natural structures or with high risk potentials.

Safety and health of customers are further important target quantities. This concerns the accident-proof construction of consumer and industrial goods, but apart from that also a large number of less spectacular product effects. So a preventative complete declaration of food ingredients or textile constituents can protect allergic persons. A careful selection of those constituents can help to avoid long-term health damages. It has been found, however (IMUG 1995), that it appears that for most companies in the food sector today it is still relatively unimportant to consider consumer health and interests.

What can a company do? Examples of tangible goals and measures:

- reduce industrial accidents per 100 000 working hours (Lost-Time-Accident frequency rate)
- use only error-friendly production technologies
- declare all constituents of consumer products

Guidelines for Economic Sustainability

Time and again, the economy points at the fact that economic sustainability is the most important and most difficult challenge in doing sustainable business. But it remains unanswered the difference between a sustainable and a non-sustainable economy.

By stressing the importance of economic goals in terms of real goods, e.g. in the context of an ecological company management, Stahlmann expected monetary quantities to increasingly lose their function as guides and signals in the macroeconomic as well as the microeconomic area (Stahlmann 1994). Here, like with many other authors, we find the hope of ecological economics to achieve a more holistic company management via an increased emphasis on ecological (material) and social items. Nevertheless it seems necessary to study also the question of how a company's value added is distributed (a question that might as well be subsumed to the social field).

The social report 1992/93 for the German media group 'Bertelsmann' goes even further. It presents the distribution of the company's financial performance among suppliers (purchase costs), government (taxes), labour (wages and salaries, social insurance), and capital (dividends).

The nowadays fashionable shareholder value concept demands maximization of profits at the cost of the three other categories in the presented distribution scheme. "The social responsibility of business is to increase profits (and nothing else!)" formulated Milton Friedman in 1970 (quoted from Ulrich 1993). With a profit rate of 100% of its turnover, an enterprise is not operable, or is no longer an enterprise – this is trivial. Hence there has to be something like a fair distribution of the company performance, fair prices for purchases of goods and services, fair wages, and a fair public share. Only such a fair distribution will, in the long term, secure a company's existence, preserve the resource of human labour, and guarantee the social peace of the society in which the company is embedded. This was the approach of a social report, where a value-added statement was the most frequently applied element. For example, for years the German weekly "Die Zeit" assessed transparency of distribution of the value added as so important that it published the value-added statements of the 100 biggest German companies (Hemmer 1996).

Table 1.   Distribution of the Bertelsmann company performance
Source: Bertelsmann social report (1992, 1993)

distribution of the company performance 1992/93 (in Germany and abroad)

	million DM	percentage	
advances			**advances**
including depreciations as advanced payments with a period lag, and bank interests (after-tax each)	10 489.2	54.9	13 399.1 million DM (70.1%)
public share			
a) taxes on advances	2 909.9	15.2	
b) taxes on Bertelsmann's value added (taxes according to profit and loss account, wage taxes of employees, value-added tax)	1 745.4	9.1	
staff			**value added**
wages and salaries, welfare charges, old-age pension scheme and relief, profit sharing (minus wage taxes each)	3 405.8	17.9	5 703.7 million DM (29.9%)
shareholders/participating certificate holders			
participation certificate dividend minus taxes	90.3	0.5	
shareholders and individual partners (share of annual profit minus shareholders' personal income taxes and capital taxes)	462.2	2.4	
total company performance			
(turnover 17.2 billion DM + further returns/turnover taxes 1.9 billion DM)	19 102.8	100.0	

Today the prevailing neo-liberal shareholder value concept is no longer sustainable in that manner. Instead, it is a short-term enforcement of individual interests at the expense of all actors who are less powerful. Obviously, the neoliberal economic system leads thus to increasing economic disadvantages for the third world, for women, and for low-qualification workers. Economic sustainability has to strive for the counterbalancing of such disadvantages instead of increasing them.

In this context, the sustainability concept could be interpreted in the sense of an intergenerational and intragenerational justice, that is, the just distribution of (life) chances among those that live today as well as among present and future generations. This yields the claim for justice of economic distribution. The following two guidelines attempt to discuss the issue of distribution justice at the national and international level.

Guideline 1:"Guaranteeing Basic Needs – Orienting by the Necessities of Life"

It is the basic purpose of all economic activities to satisfy social needs. Hence entrepreneurial activity should also orient by "satisfaction of needs". However classical economics knows (almost) nothing but the market as a gauge to judge performance with respect to this criterion. If a product is successfully placed on the market, apparently it satisfies existing needs.

It should, however, rather be demanded that a functioning economy guarantees the satisfaction of basic needs, or, in other words, the "necessities of life" for all people. This does not only mean production of products but also includes the creation of jobs so that those goods can be purchased with earned money.

One focal question is this: what do we need for (a good) life? The answer comprises necessities of a material nature (which have been the main subject of the sustainability debate so far, and which have mostly been confined to solely commercial products), but also necessities of care (be washed, nurtured, have one's provisions refilled, have one's room tidied, be informed …), and emotional or immaterial necessities (be sheltered, have friends, do a meaningful and satisfying work, know that one's children are educated and guided with loving care, be well-integrated even if being old and no longer efficient …).

What needs are assessed as necessary for life? There can be no prefabricated answers to this question. We have to reflect on our own needs, discuss them, and not just see them as already existing and therefore to be satisfied. Such a discursive consideration will lead us to take our own needs – and thereby our own nature – seriously and so not too quickly equal the purchase of goods and services with the satisfaction of needs. These discussions are important furthermore as they reveal that even the orientation by essentials of life is something culturally embedded (Jochimsen et al. 1994).

The right to welfare, the right to work, to appropriate work conditions, to equal wages for equal labour, and moreover to food, clothing, housing, and property are laid down in the General Declaration of Human Rights as economic-socio-cultural rights. But they are much less respected than civil rights. Their realisation depends to a far higher degree on the actor economic enterprise.

Companies enable or prevent the satisfaction of human basic needs by means of their products and especially through their practices of buying and selling. Sure the production of luxury goods for the rich is often market-adequate. But in view of basic rights it has to be criticised where it competes for raw materials with the production of necessary basic products.

With respect to the right to work, again companies are the crucial player. It goes increasingly undisputed that companies (and even government institutions) equal job cutbacks with lower costs or higher profitability. More and more, the entrepreneurial element to develop, on the basis of human qualification and labour, new organisational structures, new projects, or new products, and thus create jobs instead of shedding them, recedes into the background. And yet, the individual situation is not at all as clear-cut as that. Sometimes investment costs for automation are so high already today that job shedding is no longer profitable. For example, the efficiency calculation of a pharmaceutical firm in Southern Germany with regard to a new store led to a decision against automatic high-rise storage and in favour of human labour – for reasons of cost-effectiveness.

What can a company do? Examples of tangible goals and measures:

- preferably develop and produce products that satisfy necessities of life
- check any investment for its potential to preserve existing jobs or create new ones with costs
- remaining the same
- introduce a regulation about voluntary temporary part-time work
- introduce part-time jobs in management positions

Guideline 2:"Promoting the Third-world Countries"

A general consensus exists on the justifiable claim of people in the Third World to participate in global prosperity. But few propositions are made about how they could enjoy their justified share. Nevertheless, there are a whole series of reasons why even the countries of the Northern hemisphere could develop an interest in the welfare of third-world countries. For instance, we want to have our supply of raw materials from these countries guaranteed in the long run. We want to maintain them as markets for our products in the long run. And after all, we want the flow of refugees from these countries not to grow indefinitely. A realisation of these wishes presupposes socially, ecologically, and economically stable societies in third-world countries.

At the same time, our companies and our national economies neither orient their actions by a realisation nor even by a promotion of such stable societies. Fair trade is hardly a relevant issue for companies. The solution of the Third World debt crisis is not really on government agendas. The refugee problem will probably only temporarily be contained by immigration laws and deportations. It is easy to put the problems of the Third World out of our minds: they are so far away. And this has not at all changed today in the time of the global communication society. It is essentially due to the anonymity of global trade. The coal in our thermal power stations, the iron ore, the bauxite, the peanuts, the rubber, and the bananas: hardly anybody knows where they finally come from. The London Commodity Exchange defines the prices, the commodities arrive anonymously in Europe.

International buying, particularly the import of raw materials from developing countries, is thus often done at prices that do not allow the producers to lead a life in human dignity. Fair trade tries to practice a new modus vivendi here, which, apart from individual economic actors' justified strive for profits, aims at ensuring a "fair share" to all producers and traders involved in a certain product life cycle.

The mail-order firm 'Team' suggests the following criteria for a fair price: it has to cover:

- costs of production, processing and selling,
- costs of living (e.g. food, clothing, housing, medical care, education),
- costs of investments into the future (e.g. improvement of production methods, product diversification to decrease dependencies).

A crucial precondition for fair trade is direct co-operation, knowledge about trade partners, and a conscious abolition of the anonymity of raw material markets.

What can a company do? Examples of tangible goals and measures:

- find out which raw materials it receives from third-world countries
- provide these raw materials on the basis of fair trade or from cooperative projects
- support third-world countries on request with the transfer of know-how
- build up specific partnerships with third-world actors
- finance business with third-world countries in a way that does not aggravate their debt crisis

---

The Path to the Sustainable Company: a Change of Values is Indispensable

The goals of the Environmental Management and Audit Scheme orient firstly by environmental policy and only then by company policy. It can easily be seen, however, that orientation by the above-listed paradigms of ecological, social, and economic sustainability goes beyond the framework of environmental policies established by companies. These are usually restricted to some paradigms of environmental protection as well as (partial aspects of) safety and health.

So we have to observe that an orientation by the system of objectives presented here does not simply mean a continuation of the existing improvement process, it will require a radical change of a company's policy.

We have to do away with a whole series of models of classical economics:

- Company policy must no longer orient merely by the 'homo economicus', this remarkably clever guy with his somehow one-sided gifts, who lacks any ability for critical self-reflection (Ulrich 1993), but instead also by partnership and co-operation.

- Orientation by shareholder values has to be complemented by an orientation by stakeholder values.
- Globalisation of markets includes not only potentials but also responsibilities to a much greater extent. This has to be understood and integrated into company policy.
- The largely undisputed conviction that any shedding of jobs will be useful for a company's profitability has to be replaced by an active responsibility for jobs.
- The idea that a company could become ready for the future by a little bit of environmental management has to be replaced by a holistic orientation by ecological, social, and economic sustainability.

Of course, a good practice of defining goals, of controlling them, and of reporting about progress, is important for sustainable companies as well. They have to consequently transfer their experiences made by means of environmental management to the fields of new objectives. But how can such new objectives be defined? How can those areas be identified where the increasingly complex economic, ecological, and social demands meet with profitable, and hence for a company, sensible ways of doing ethically responsible business (Ulrich/Fluri 1992)?

The Action Principles Transparency, Dialogue, and Co-operation

Almost all management approaches of the last 15 years – from participative management via organisational development to company culture – postulate more or less explicitly a "change of paradigms" towards communicative integration and co-operation forms (Ulrich 1993). Starting from the assumption that individual economic interests only flourish in a friendly environment, the task is to identify negative external effects and the social groups concerned (stakeholders), and, as far as possible, develop entrepreneurial solutions for existing problems together with stakeholders. A company policy orientated by a consensus with people concerned, respecting them as mature persons, deciding together with them, and seeking to balance interests, has the best chance of removing the causes of social conflicts.

Such a consensus-oriented company policy is based on the conviction of practical reason that contradictions among the above-mentioned economic, ecological and social categories cannot be removed by the entrepreneur alone after considering interests in a paternalistic way. Instead, there are continuously changing equilibria in different interest constellations, which can only be identified or promoted by means of dialogue.

Co-operation needs information, understanding, and the attainment of a consensus and therefore requires qualities like the ability to communicate, flexibility, tolerance, and the ability of self-organization (qualities that are time and again attributed to women). Co-operation thus presupposes transparency and frankness in dialogue. And co-operation partners have to be informed and trained so as to have at their disposal the necessary knowledge for a successful co-operation.

The respective attitudes and ways of acting have to become an integral part of life: act in solidarity, come to a mutual understanding, be ready to attain a consensus, be open to arguments, listen with active participation, be able to put aside short-term and medium-term self-interests in favour of long-term common interests, take time for others and for problems, be patient, keep things, experiences and relationships, act carefully, follow reason and not hierarchy, be understanding, etc. The demand to act in such a way requires us to put in question the concept of the "innately egoistic" human being, the homo economicus (Biesecker 1994, and Ulrich 1993).

Companies are especially affected by the demand for information about impacts, backgrounds, and plans regarding their activities. The reporting about economic issues is largely regulated by law. The approach of the 70's to publish social reports has essentially failed. But there have been promising developments with environmental reporting since the beginning of the 90's (Clausen/Fichter 1996 and 1998). These should be extended to a sustainability reporting within the years to come.

What can a company do? Examples of tangible goals and measures:

- cooperate in the long term with suppliers (development partnership, long-term supply contracts) and customers (specialized dealers)
- cooperate in the field of material flow management
- inform openly, e.g. by means of company, environmental, and social reports
- declare completely the constituents of products towards customers and, possibly, final consumers
- carry on a dialogue with all important stakeholder groups
- make employees participate as far as possible

Conclusion

Important for the sustainable company is the change of values in favour of a sustainable company policy with economic, ecological, and social goals. This inevitably means to break with some prevailing and familiar dogmas. It is recommended to seek the frank dialogue with all groups indirectly or directly concerned by the company's activities and to look, together with them, for areas where entrepreneurial success meets with ethically responsible action (in the context of the sustainability concept). Only in that way will it be possible to understand new goals not merely as a process of overcharging companies with social demands, but as a multitude of chances. These will be open in future to those companies which are the quickest to identify existing synergies, market potentials, and new possibilities. Each challenge in the form of a changing political environment can be used by companies that flexibly and undogmatically realise innovative solutions instead of grieving over the loss of the old.

References

Bertelsmann AG (1993): Sozialbilanz 1992/93, Gütersloh.

Biesecker, A. (1994): Wir sind nicht zur Konkurrenz verdammt, In: Busch-Lüty, C.; Jochimsen, M. et al.: Vorsorgendes Wirtschaften; Politische Ökologie special issue 6; Munich, p. 28–31.

Busch-Lüty, C.; Jochimsen, M. et al. (1994): Vorsorgendes Wirtschaften; Politische Ökologie special issue 6; Munich.

Clausen, J.; Fichter, K. (1996): Umweltbericht – Umwelterklärung: Praxis glaubwürdiger Kommunikation von Unternehmen, Carl Hanser, Munich.

Clausen, J.; Fichter, K. (1998): Environmental Reports – Environmental Statements – Guidelines on Preparation and Dissemination, INEM/ future e.V., Munich, Wedel.

Council on Economic Priorities Accreditation Agency (1997): Social Accountability 8000, London.

Fischer-Kowalski, M. et al. (1993): Das System verursacherbezogener Umweltindikatoren, Schriftenreihe des Instituts für ökologische Wirtschaftsforschung No. 63/93, Berlin.

Frenssen, G. (1903): Die drei Getreuen, Grote´sche Verlagsbuchhandlung, Berlin.

Gleich, A. von; Hellenbrandt, S. (1994): Effizienz und Risiko; IÖW-VÖW Informationsdienst 9 No.5.

Gleich, A. von; Rubik, F. (1996): Umwelteinflüsse neuer Werkstoffe, VDI Fortschrittsberichte, series 15, Düsseldorf.

Hemmer, E. (1995): Das Scheitern einer gescheiten Idee in: Arbeitgeber 23/48 pp. 796–800.

Institut für Markt, Umwelt und Gesellschaft (IMUG) (1995): Der Unternehmenstester – Die Lebensmittelbranche, rororo, Reinbek.

Jochimsen, M.; Knobloch, U.; Seidl, I. (1994): Vorsorgendes Wirtschaften – Konturenskizze zu Inhalt und Methode einer ökologischen und sozialverträglichen Ökonomie in: Busch-Lüty et al., p. 6–11.

Stahlmann, V. (1994): Umweltverantwortliche Unternehmensführung, C.H. Beck, Munich.

Ulrich, P. (1993): Integrative Wirtschafts- und Unternehmensethik – ein Rahmenkonzept, Beiträge und Berichte No. 55 of the university St. Gallen.

Ulrich, P.; Fluri, E. (1992): Management – eine konzentrierte Einführung, 6th reviewed and complemented edition, Bern/ Stuttgart.

UNO (1948): General Declaration of Human Rights, New York.

3 The Influence of Environmental Regulation on Company Competitiveness: A Review of the Literature and Some Case Study Evidence

David M.W.N. Hitchens

Introduction

This chapter is concerned with the effect of environmental regulations on company competitiveness. There are two parts, the first provides an overview of the literature on the general question as to what is the relationship between environmental regulations and economic performance. The second reports the findings of a study of the effect of environmental regulation on the competitiveness of food processing firms across four countries and six regions in the EU. Firms located in these regions face different regulatory policies and it is argued have differing capabilities to absorb environmental compliance costs.

Part 1 Does a Stringent Environmental Policy put Firms at a Competitive Disadvantage?

Do environmental regulations impact on the competitiveness of firms? Environmental policy requires that the firm redirects resources from other profitable opportunities which can lead to a rise in costs and prices and the loss of markets. This is because the basis of environmental policy is the polluter pays principle (PPP) (OECD 1975). The polluter bears the expense of carrying out measures to prevent and control pollution. There is an alternative view, however, and this is that environmental policy improves competitiveness by pushing firms into developing more efficient ways to produce and therefore reduces costs. Some would go so far as to argue that stringent environmental policy is a potent form of industrial policy, that it provides a double dividend whereby it both improves the environment and competitiveness (Simpson and Bradford 1996).

How Important are Environmental Compliance Costs?

How important are environmental costs? Environmental compliance costs are usually measured to be small. Across manufacturing they account for well under 1 per cent of gross output but vary across industries with the highest costs incurred in the primary metals, non metallic mineral products, chemicals, paper and food industries (DOE 1996; US Department of Commerce 1993; Sprenger 1996). Costs include investment in capital plant and equipment, increased operating costs, R&D to reduce costs, administrative and legal costs. However, only the first two are usually counted. Even direct costs arising from investment in clean technologies and changes to production processes and end products are excluded because of difficulty in identifying and measuring these. Other costs include uncertainty created by detailed command and control, with respect to new products and processes and the consequent reduction in rates of return through discouraged innovation and slow growth. Measured costs only include increased operating costs and investment following regulation.

Defining Competitiveness

Most economists would emphasise the need to judge the competitiveness impact of environmental regulation by measuring the effect on productivity (though there is much debate (Krugman 1996)). For example, the OECD (1992) defines a nation's competitiveness as 'the degree to which it can, under free and fair market conditions, produce goods and services which meet the test of international markets, while simultaneously maintaining and expanding the incomes of its people over the longer term'. The UKs White Paper on competitiveness (1994) defines competitiveness 'as about creating a high skills, high productivity and therefore a high wage economy'. The European Commission White Paper (EC 1994) also bases its notion of competitiveness on productivity. While there have been studies of the impact on productivity of environmental regulation (and some of these are considered below), common macro economic approaches focus on the effects on trade and foreign investment (OECD 1993).

What indicators are there of the competitiveness performance of firms? There are output side indicators (i.e. those which illustrate the consequences of the relative competitiveness of the firm): profitability, market share, productivity, patents etc. There are also a set of indicators which correspond to the input side (i.e. representative of the likely explanations of competitiveness): physical and human capital, R and D spending, stock to turnover ratios etc. (Jacobson and Andréosso-O'Callaghan 1996). The importance of these indicators is considered in more detail in Part 2.

The Macroeconomic Evidence: The Impact on Trade Flows and Foreign Direct Investment

One focus has examined the effects of regulation on trade to test for a loss of comparative advantage in environmentally sensitive industries. The question addressed is whether highly regulated industries suffer in terms of exports, whether production moves abroad and whether there is increasing investment by firms overseas (Low 1992). Less attention has been paid to that which actually trades (Oosterhuis et al. 1996) although stringent regulations can prohibit the trade of non complying imports. For example German beer producers claimed that they could not compete in Denmark because of Danish rules on the use of appropriate containers for recycling.

Research concerned with trade effects examines whether environmental regulations impose significant costs and reduce exports particularly in the pollution intensive sectors and whether international patterns of trade have been effected by inter country differences in regulation (Kalt 1988; Tobey 1990; Sorsa 1994). The results of the studies have been largely inconclusive, and failed to find a significant relation between environmental regulation and trade performance. One of the main reasons is that even when trade patterns are shown to change there is no shortage of competing explanations for the phenomenon.

Another concern at the *macro* level has been with industrial flight (Leonard 1988; Low and Yeats 1992; Lucas, Wheeler and Hettige 1992; Olewiler 1994; Repetto 1995). These studies focus on the migration of polluting industries from countries with strict regulations. If environmental regulations are having a significant impact on firms they will be investing more abroad. The findings of these studies are mixed, and where the research shows that dirty industries have migrated, the authors have found it very difficult to conclude that the dispersion is due to environmental policy.

Microeconomic Evidence: The Debate Regarding the Porter Hypothesis

It may be that the significance of the impact of environmental regulation can be more easily understood from a study of firms and industries. There are a number of factors which are important. In general terms a negative impact on the output and employment of firms will be greater the greater the rise in costs following compliance; the greater the differential cost penalty relative to domestic and foreign competitors; the more significant the costs are in total costs; the greater the degree of price competition between firms and the greater the sensitivity of demand to price increases (OECD 1993).

On the other hand industries which are characterised by higher rates of investment may be able to take advantage of cost reducing clean technologies. Consumer preferences may shift in favour of green products and cleaner production. Regulation may stimulate innovation and raise productivity, if policy provides the appropriate incentive.

The importance of the balance of these negative and positive factors has given rise to what is known as the *revisionist* view of the importance of regulation, and is associated with Michael Porter (Porter 1990, 1991). He hypothesises that innovation takes place in response to regulation and leads to *innovation offsets* (Porter and van der Linde 1995): "innovation addresses environmental impacts while simultaneously improving the affected product itself and/or related processes...in some cases these innovation offsets can exceed the costs of compliance".

And, "*companies simply get smarter about how to deal with pollution..*". Regulation signals resource inefficiencies by focusing on information gathering, for example, where companies are inexperienced about measuring their discharges.

While green products and environmental technology provide opportunities for, "*early mover advantages in international markets*", for those companies which innovate earlier than in other nations.

The background to these innovation offsets lies in technological solutions to environmental problems, especially clean technologies. It is difficult to define clean technologies but in general terms they involve both processes and products to minimise waste, reduce materials and energy inputs, minimise the overall environmental impact at all stages of production and consumption. A step change to cleaner technology involves the initial cost of redesigning processes and products, but it reduces waste, and can lead to offsets, in contrast to 'clean-up' or end of pipe technology (Clift 1993).

To deliver the favourable outcome Porter stresses the need for good environmental measures, i.e. those which use market incentives, taxes and subsidies, and stress innovation and adoption of clean production methods. Compared with poor regulations, based on command and control, which constrain the choice of technologies and stress end of pipe solutions and clean up measures. So what is critical is that the firm has technological freedom to choose the compliance technique, though consultation between government and industry is useful where the regulatory body is aware of the most recent advances in industrial technology (Wallace 1995).

Porter points to the productivity performance of the Japanese and German economies as 'the strongest proof that environmental protection does not harm competitiveness'. In fact a substantial amount of environmental innovation does occur and there is strong empirical evidence of a relationship between pollution abatement expenditures and the patenting of environmental technologies in developed countries (Lanjouw and Mody 1996). The lead is taken by Germany, USA and Japan, the countries with the most stringent regulations, in all areas of environmental technology. These countries also have the largest industries and markets. However the scope for first mover advantages in the environmental technology fields is limited since trade is hampered by a lack of harmonisation of policies, so the eco industry is not producing standardised products and importing firms have to at least adapt technology to their own country's regulations (Palmer and Simpson 1993; Palmer, Oates and Portney 1995; OECD 1996).

Economists are Sceptical

Economists view with enormous scepticism the idea that businessmen systematically overlook profitable opportunities. They would argue that although innovation offsets are possible they are likely to be small and that systematic under research by firms is unlikely to be the case. In addition they argue that the case study evidence cited to support the hypothesis is particular and not representative.

While economists acknowledge that firms are unlikely to work to maximum efficiency and there exist models of management behavior which incorporate satisfactory performance as an organizational goal of the firm, the idea that there is widespread slack, where management forgoes opportunities to discover cost reducing R&D is strongly doubted. They also doubt that where slack does exist that environmental policy is the right policy through which to improve efficiency.

Case Study and Productivity-related Evidence

Most evidence is based on case studies. These case studies aim to show that a wide variety of innovations are occurring and lead to competitive advantages, either arising in the regulated industry itself or the supplier industry. Successful examples abound many published by the US EPA (Management Institute for Environment and Business 1996). A recent UK example (ACOST 1992), quotes how stringent restrictions on airport noise in the UK created markets in the US for aeroengines developed by Rolls Royce, because these were significantly quieter than those of their rivals. The regulations created the same advantage for the 'Whisper Jet'. But these case studies provide insufficient evidence on which to base policy. More systematic research evidence is required.

Three studies are of note. A paper by Gray and Shadbegian (1995), based on a study of plants in three industries in the US (pulp and paper, oil refineries and steel mills) found a negative relationship between a plant's pollution abatement costs, i.e. the extent to which it was regulated, and productivity. The heavier the pollution and regulation, the lower the productivity. Raising environmental performance in these industries did not induce productivity benefits large enough to outweigh the measured compliance costs.

Another study by Barbara and McConnell (1990) reports on a study of five heavily polluting US industries. They attempted to measure some of the positive impacts of regulation. They measured the direct effects of regulation, these are the direct pollution abatement costs, and these always have a negative productivity effect, because they add to total input costs. They also measured indirect effects, the change in inputs required to manufacture the core products of the business. This has the potential for offsets.

In only two of the five industries examined were any positive indirect effects found (these were in the case of non ferrous metals over the entire time period covered and in the chemicals industries over the period 1960–75) and these were insufficient to offset the direct compliance costs. In the other three industries not only did the direct costs of pollution abatement add to total costs but there was also an increase in indirect costs. Rather than finding any offsetting of costs they found there was a need to increase the inputs to the conventional manufacturing process in addition to the required spending on pollution abatement.

Finally, a study undertaken by Repetto (1995) tested the hypothesis that superior environmental performance tends to lead to lower profitability than inferior environmental performance, within the same industry. This is the standard hyphothesis, better environmental performance comes at a cost, so that firms need to divert resources to reduce their emissions and must sacrifice profits. He tests this hypothesis against the Porter hypothesis, that firms motivated to find solutions to environmental problems, by regulation, find previously overlooked cost saving opportunities to improve processes, reduce waste or redesign products. In particular he tests the form of the Porter hypothesis that firms with superior environmental performance also achieve superior profitability within their industries.

The findings are based on 50 finely defined industries. Environmental performance is measured in physical terms e.g. in the case of water pollution using , biological oxygen demand and suspended solids per dollar of sales, against profit on sales and profit on net assets. He found that there was no overall tendency for plants with superior environmental performance to be less profitable. He concluded that there is absolutely no evidence that superior environmental performance puts firms at a market advantage or that it adversely effects market performance.

Summary of the Evidence

To summarise, research shows little evidence to support the hypothesis that environmental policy leads to a loss of competitiveness as measured by a reduction in trade or industrial flight to pollution havens. There is also a lack of systematic evidence to support the Porter view that environmental regulation improves competitiveness through 'innovation offsets' and/or 'front runner' advantages.

Part 2   Impact of Environmental Regulations on Firms in a Peripheral Region in the EU

This section reports on a study[1] to determine the effect on competitiveness of the costs of environmental compliance with respect to waste water and solid waste in

---

[1] I would like to acknowledge the research collaboration of Esmond Birnie and Angela McGowan (Queen's University, Belfast) and Ursula Triebswetter (IFO, Munich) and A Cottica (Eco and Eco, Bologna). The study was undertaken for the European Foundation (Dublin) (Hitchens et al. 1996). Help and support were received from Charles Robson (European Foundation) and Bill Watts (European Commission, DGXII/D-5).

the dairy and meat processing industries in response to environmental regulation across four countries and six regions in the EU. By a careful matching of plants by size and product type between countries and regions where firms face different regulatory costs and achieve different levels of competitiveness, it was possible to isolate the effect on firm competitiveness of the cost of compliance.

Samples of firms were drawn from Germany, Northern Ireland, the Republic of Ireland (because of similarities in economic performance and environmental stringencies facing firms in the two Irish regions sample results for the two economies – "Ireland" – were pooled) and Italy. The choice of Germany reflected longstanding high environmental standards, and included firms located in East Germany where a recent and sudden change from low standards towards the higher West German standards has occurred. In contrast, the Irish economies, and north and south Italy have less rigorous standards together with a perception of weaker enforcement.

The object of the research was to investigate (1.) whether those firms located in regions where manufacturing productivity was typically weak were at a disadvantage relative to firms in stronger regions in adjusting to regulation and (2.) whether firms could achieve international and national standards of competitive performance while facing different regulatory stringencies and/or incurring different environmental compliance costs.

First the regulatory background in each country is discussed. This is followed by a consideration of the differences in the productivity performance of firms between each of the selected regions, and a discussion of why firms in low productivity regions may find it more difficult to adjust competitively to regulation.

Environmental Waste and Regulatory Background

The two main sources of waste in dairy and meat processing are solid waste, typically in the form of packaging, animal waste etc. and liquid waste. The principal environmental impact arises from liquid effluent (DTI 1991), and the main environmental consideration is with the disposal of that waste into sewers, rivers, streams, lakes or estuaries.

There are licence or permit or consent systems whereby the authorities seek to control the total amount of waste water discharge by setting a maximum permissible level of discharge by restricting the right of individual firms to discharge within this limit. The quality of discharge of individual plants depends particularly on whether the plant is discharging to rivers, streams etc. in which case the effluent must be treated by the firm to a certain standard, or whether discharge is to sewers in which case the local authority undertakes the treatment to the required standard. Direct dischargers treat sewage within their own plant and can then dispose of it into rivers, estuaries etc. Indirect dischargers are those companies which rely mainly on the sewerage system to treat their waste (though there may be requirements to pretreat waste water to remove specified dangerous substances).

In Germany and Italy a national standard is set for direct dischargers, while in Ireland the standard varies and individual firms are licensed according to the nature of the environment into which the discharge will occur, e.g. whether the river is a stream or an estuary, fast flowing or slow flowing. Irish standards, because of this geographical variation, can be either stricter or more lax than the standards set for counterpart Italian or German firms. The strictness of the ultimate standard affects the cost of cleaning waste.

In Germany charges are levied on direct dischargers (i.e. those firms discharging to waters; rivers, streams, estuaries etc.) and indirect dischargers (i.e. those discharging effluent to the sewerage system where it is subsequently treated), with charges being set on a regional basis. In Italy and the two parts of Ireland charges are levied *only* on indirect dischargers. In Italy charges are set at a regional level (with some coefficients allowed to vary to reflect economies of scale in treatment plants) and in the Republic of Ireland charges are set by the local authority. In Northern Ireland the charging system, operative since 1992, is centrally organised. The German system has the toughest standards and has been operating for the longest period and involves the highest per unit charges.

Comparative Productivity Performance

Official production census data show that Italian dairy and meat processing firms achieved higher productivity than their German and Republic of Ireland counterparts in the early 1990s. Comparisons of total labour costs per employee (using exchange rates) confirm the rank ordering provided by the productivity comparisons. Republic of Ireland firms have the poorest productivity performance. These census relationships were also reflected in the sample data, and they therefore provide corroborative evidence of sample representativeness. Turning to regional differences: productivity levels in East German food processing has been estimated at about 40 per cent of the West German level in 1987 (van Ark 1994), and has still not converged on the west German level. In the two parts of Ireland productivity in indigenous industry is estimated to be similar at between 80–90 per cent of the UK average. In southern Italy manufacturing productivity is 70 per cent of the north Italian level.

Vulnerability of Low Productivity Regions

Firms sampled in the two parts of Ireland, east Germany and southern Italy are located in areas where manufacturing registers relatively low productivity. A poor productivity record is expected to weaken competitiveness in the face of environmental regulation because firms and countries with higher levels of productivity (competitiveness), have management and other capabilities including, work force skills, R&D efforts, up to date equipment and methods of production to more readily absorb compliance costs. As discussed in part 1 above, the competitive performance of firms may be defined from the input or output side. Where competitiveness is defined from the input side (i.e. representative of the likely explanations of productivity differences): the measure is based on physical and human capital endowment, R and D spending etc.. These are also factors which are known to influence the adoption of cleaner production technologies (Green et al. 1994; OECD 1985, 1987, 1995; ECMT/OECD 1994; Wallace 1995). Furthermore the mere fact that the firm is competitive means it has a management capability to respond to environmental pressures with best practice solutions. Output side indicators of competitive performance: profitability, market share, productivity, patents, firm growth etc., not only measure (in principle) the consequences of the adoption of clean technologies (ENDS 1994; CBI 1994; OECD 1987, 1995; Porter 1990), but also provide the resources and opportunities for the adoption of cleaner production methods and products.

Environmental Waste Costs

In this section the cost of compliance by firms in each of the three countries is reported and consideration is given to research objective (1.) to examine whether environmental costs are greater for firms in low productivity regions.

Data were collected on the cost levied for discharge to sewers and the costs incurred in self treatment. Sometimes companies use a combination of methods and discharge partially treated effluent. Data were then collected on both the cost of 'self-treatment' and the sewerage charge. In the sample, 71% of German firms discharged to sewers compared with 36% and 40% of Irish and Italian firms respectively. These percentages reflect national differences in the preponderance of indirect dischargers.

In addition to waste water costs, the cost of solid waste disposal was estimated. Solid waste in dairying and meat processing arises in the form of sludge which occurs during the course of waste treatment, animal waste e.g. stomach contents sent to landfill sites. It also arises as packaging in the form of cardboard, paper, plastics. Costs of disposal vary according to whether or not secondary processing takes place, whether waste is taken away free, for example, by a recycling contractor, or, as in many cases, compressed and collected in skips and taken to a landfill site (there were also examples where packaging waste and carcasses were incinerated). Data were collected on the cost of solid waste collection and disposal, including an estimate for labour costs where waste separation was undertaken. A number of firms purchased compactors to reduce the volume of waste and the annualised cost of these small capital items was included. Overall, and in most cases, these costs were small. Waste water costs per m^3 dominate total environmental waste costs.

Table 1 shows the distribution of the estimates of environmental costs as a percentage of turnover for the dairy and meat processing firms sampled. Total envi-

Table 1. Distribution of total environmental costs by numbers of companies

Waste Costs as a % of Turnover	East Germany	West Germany	Ireland	North Italy	South Italy
1% +	5	2	2	1	–
0.5–0.99%	2	3	4	1	2
less 0.5%	4	5	20	8	8
Sample Average	1.14	0.85	0.3	0.3	0.27

ronmental costs were (on average) around 1 per cent of turnover for the German plants (reflecting the higher charges and more stringent regulation there) and for Irish and Italian firms these were lower and varied (on average) between 0.27 per cent and 0.3 per cent of turnover. Low productivity east German plants incurred higher compliance costs than their western counterparts but compliance costs at firms in low productivity Ireland and southern Italy were remarkably similar to those of their counterparts in high productivity northern Italy. The hypothesis that firms in low productivity regions are vulnerable to environmental policy is not confirmed. The table shows that firms in low productivity regions do not necessarily incur higher environmental compliance costs.

Relative Competitive Performance and Environmental Costs

In this section the second research objective is examined. The hypothesis that firm competitiveness is negatively related to compliance costs is tested. Is there a trade off between above average competitive performance and the cost of environmental compliance?

Measurement of Firm Competitiveness

i. *Comparative international productivity performance.* Although average productivity of firms sampled differs between regions and countries, a number of firms performed to international standards in each country, as measured by their productivity performance. For example in Ireland three firms (one in Northern Ireland and two in the Republic of Ireland) recorded a high productivity performance against counterpart companies in both Germany and Italy. On the other hand, although Italian firms scored highly *on average* against their counterparts in the other countries, only four out of ten were manufacturing at a clearly high standard when set beside counterparts in Germany and Ireland. Firms which acheved high levels of *international* comparative productivity were identified in each country.

Table 2.   Environmental costs (as % of turnover) for above average performance firms

firms with above national average performance on each criterion	Germany dairy	Germany meat	Ireland dairy	Ireland meat	Italy dairy	Italy meat
International comparative productivity	0.7	1.3	0.1	0.4	0.2	0.3
Value added per head	0.9	1.1	0.1	0.3	0.2	0.4
Physical productivity per head	1.6	1.1	0.4	0.3	0.3	0.3
Exports	0.7	0.4	0.2	0.6	0.2	0.3
Growth	1.4	1.7	0.1	0.2	0.3	0.6
Sample average	0.9	1.0	0.3	0.4	0.3	0.3

ii. Relative national productivity performance. Sample firms were also subdivided *nationally* according to whether they recorded above average value added per employee and above average physical productivity per employee compared with other firms sampled in the country. Those firms with value added per employee minimally 25% above the sample average were defined as having above average competitive performance. Similarly high physical productivity (based on milk and meat processed per person) is defined as where plant productivity is 25% and above that of the national sample average.

iii. Export performance. The extent to which firms export can be taken as further evidence of the extent of competitiveness, and strong exporters were defined as exporting 25% and above the country sample average. For Irish firms in particular export markets were especially important, German dairy firms exported about 20% of their output while Italian food and German meat firms exported under 10%.

iv. Firm growth. One further indicator of competitiveness has been adopted and that is one based on employment change over the period 1990–95. Competitive firms were defined as those growing in the period.

Table 2 shows the compliance costs incurred by those firms with above average 'competitiveness performance' as implied by these measures. When their compliance costs are compared with those achieved by sample companies as a whole, shown in the final row of the table, it can be seen that above average firm performance, based on these different measures, can be achieved with above average compliance costs. In nearly half the comparisons better performing firms have average or higher compliance costs.

The results show that in all three countries firms were able to achieve above average competitive performance, on any of these measures, irrespective of the magnitude of the environmental costs they faced.

Industry Competition and Major Factors Influencing Competitiveness

Sample firms faced strong competitive pressures. The diversity of growth experience showed that firms within countries/regions were in active competition. Many firms were important exporters (two-thirds were exporting mainly to EU markets but also to world-wide markets) they therefore faced *international competition*. The principal customers were similar, sample companies served three main customers – wholesalers, retailers and other food processors – the most important of these was the retail market (dominated by supermarkets). Hence firms were operating in *similar markets*. Most sample firms were therefore open to both national and *international competition and therefore to competitiveness effects of variations in environmental regulations and costs faced by firms. Against this background the competitive advantages and disadvantages recognised by firms was examined.

The main competitive advantages reported were quality, plant location and labour related factors (especially skill and cost). A range of other advantages associated with prices, costs, availability of grants and tax advantages, and those relating to the technical aspects of production were also specified. Of note were the strong claims of product quality advantages for the German and Italian firms.

On disadvantages there were a wide range of factors given amongst which pressures on price, location (for peripheral firms), labour availability, quality and labour relations were amongst the most important. Eight respondents specified environmental costs as constituting a competitive disadvantage, seven of the firms were located in East Germany, the other in West Germany. East German firms in addition to citing high environmental costs as competitive disadvantages also considered location, market overcapacity and a lack of high value added products as key factors. West German firms were more concerned about costs arising from compliance with hygiene regulations than with waste water regulations, in addition to other factors. Environmental compliance costs accounted for only 6% of disadvantages cited.

The stress on environmental costs in East Germany, arose at new plants (these comprised 60% of East German interviews). Managers at these companies had made considerable recent capital investments to meet the environmental regulations, and it was about these costs, and consequent high waste water treatment costs, that they referred. Investment in environmental protection was not large by overall plant investment requirements and was not specified as a constraint on future growth. At existing East German dairy and meat plants, sampled in Stotternheim, Schwartza, Mutzschen and Ostthuringen, no manager said that environmental legislation had proved a competitiveness or growth constraint.

Conclusions

The first part of the chapter showed that there is little evidence to support the hypothesis that environmental policy leads to a loss of comparative advantage or industrial flight to pollution havens (Jaffe et al. 1995), and there is little statistical support for the revisionist view that regulation improves competitiveness and therefore should be tightened. Hence from the point of view of the firm, the weight of evidence suggests that environmental regulation does not influence competitive performance.

The second part of this chapter reports on a matched comparison of food processing firms between Ireland, Germany and Italy. The sample included firms located in the relatively low productivity regions of east Germany, southern Italy and the two parts of Ireland. It was hypothesised that a poor productivity record could weaken competitiveness in the face of environmental regulation because of a reduced capability by such firms to absorb compliance costs. In additon there was variation in regulatory stringency and enforcement between countries and regions; Germany represented relatively stringent standards and Ireland and Italy less rigorous standards.

The research showed that firms in high productivity regions do not necessarily adjust more efficiently and therefore more cheaply to regulation, than those firms in poorer regions. Moreover the ability to compete both nationally and internationally was unaffected by small differences in the costs of compliance. Tough regulations in Germany did not prevent firms there from achieving international standards of competive performance. Other factors (besides environmental regulation) had a more important influence on a firm's competitive position both nationally and internationally.

References

Advisory Council on Science and Technology (ACOST) (1992): *Cleaner Technology*, HMSO, London.

van Ark, B. (1994): "Reassessing growth and comparative levels of performance in eastern Europe: The experience of manufacturing in Czechoslovakia and East Germany", *Paper given to the Third EACES Conference*, Budapest, Sept. 8–10.

Barbera, A.J.; McConnell, V.D. (1990): The impact of Environmental Regulations on Industry Productivity: Direct and Indirect Effects, *Journal of Environmental Economics and Management*, 18, 50–65.

CBI (1994): Environment Costs, *Confederation of British Industry*, London.

Clift, R. (1993): Pollution ad Waste Management 1: Cradle to Grave Analysis, *Science in Parliament*, 50 (3) June.

Department of the Environment (1996): Environmental protection expenditure by Industry, Department of the Environment, HMSO.

Department of Trade and Industry (1991): *Cleaner technology in the UK*, PA Consulting Group HMSO London.

EC (1994): *Growth, Competitiveness, Employment: The Challenges and Ways forward in the 21st Century*, White Paper, European Commission, Brussels.

ECMT/OECD (1994): *Internalising the Social Costs of Transport*, ECMT/Organisation for Economic Co-operation and Development, Paris.

ENDS (1994): *Integrated Pollution Control; the first three years*, Environmental data Services Ltd, London

Gray, W.B.; Shadbegian (1995): Pollution Abatement Costs, Regulation and Plant Level productivity, NBER Working Paper No 4994, NBER, Washington, DC.

Green, K. et al. (1994): Technological Trajectories and R&D for Environmental Innovation in UK firms, *Futures*, 26(10).

Hitchens, D.M.W.N.; Birnie, E.B.; McGowan, A.; Triebswetter, U.; Cottica, A. (1996): *Effects on Employment, Skills, Productivity and Competitiveness of Environmental regulation in Food Processing (Dairy and Meat Processing) across the EU (Northern Ireland and the Republic of Ireland; Germany East and West; Italy, north and south)*. European Foundation, Dublin, pp.254. Forthcoming (1998) *The Firm, Competitiveness and Environmental Regulations*, Edward Elgar.

HM Government (1994): *Competitiveness*, Cm 2563. HMSO, London.

Jacobson, D.; Andréosso-O'Callaghan, B. (1996): *Industrial Economics and Organisation*, McGraw Hill, Maidenhead.

Jaffe, A.B.; Peterson, S.R.; Portney, P.R.; Stavins, R.N. (1995): Environmental regulation and the competitiveness of US manufacturing: What does the evidence tell us?, *Journal of Economic Literature*, 33, 132–163.

Kalt, J.P. (1988): The impact of Domestic Environmental Regulatory Policies on US International Competitiveness, *International Competitiveness*, Cambridge, MA. Harper and Roe, Ballinger.

Krugman, P.R. (1996): Making Sense of the Competitiveness Debate, *Oxford Review of Economic Policy*, 12(3),17–25.

Lanjouw, O.L.; Mody, A. (1996): Innovation and the International diffusion of Environmentally Responsive Technology, *Research Policy, 25*, 554–571.

Leonard, H.J. (1988): *Pollution and the Struggle for the World Product: Multinational Corporations, Environment and International Comparative Advantage*, Cambridge University Press, Cambridge.

Lowe, P. (ed.) (1992): *International Trade and the Environment*, World Bank Discussion Paper 159, World Bank, Washington, DC.

Lowe, P.; Yeats, A. (1992): Do 'Dirty' Industries Migrate?, in Lowe, P. (ed.), *International Trade and the Environment* pp. 89–103. World Bank, Washington DC.

Lucas, R.E.B. Wheeler; Hettige, H. (1992): Economic Development, environmental regulation and the international migration of toxic industrial pollution: 1960–1988, in Lowe, P. (ed.), *International Trade and the Environment* pp. 67–86 World Bank, Washington DC.

Management Institute for Environment and Business with US Environmental protection Agency (1996): *Competitive Implications of Environmental Regulation: a study of six industries*, MIEB/US EPA, Washington, DC.

OECD (1975): *The Polluter Pays Principle*, Organisation for Economic Co-operation and Development, Paris.

OECD (1985): *Environment Policy and Technological Change*, Organisation for Economic Co-operation and Development, Paris.

OECD (1987): *The Promotion and Diffusion of Clean Technologies*, Organisation for Economic Co-operation and Development, Paris.

OECD (1992): *Technology and the Economy: The Key Relationships*. Organisation for Economic Co-operation and Development, Paris.

OECD (1993): *Environmental Policies and Industrial Competitiveness*, Organisation for Economic Co-operation and Development, Paris.

OECD (1995): *Technologies for Cleaner production and Products,* Organisation for Economic Co-operation and Development, Paris.

OECD (1996): *The Global environmental Goods and Services Industry,* Organisation for Economic Co-operation and Development, Paris.

Olewiler, N. (1994): The Impact of Environmental Regulation on Investment, in Benidickson J, Doern, B. and Olewiler, N. (Eds) *Getting the Green Light: Environmental Regulation and Investment in Canada.* C.D. Howe Institute, Toronto.

Oosterhuis, F.; Rubik, F.; Scholl, G. (1996): *Product Policy in Europe* Kluwer Academic Publishers, Dordrecht.

Palmer, K.L.; Simpson, R.D. (1993): Environmental Policy as Industrial Policy, *Resources,* Summer.

Palmer, K.; Oates, W.E.; Portney, P. (1995): Tightening Environmental Standards: The Benefit-Cost or no-cost paradigm? *Journal of Economic Perspectives,* 9(4),119–132.

Porter, M.E. (1990): *The Competitive Advantage of Nations,* Free Press, New York.

Porter, M.E. (1991): America's Green Strategy *Scientific American,* April, p168.

Porter, M.; van der Linde, C. (1995): Towards a new conception of the Environment-Competitiveness Relationship, *Journal of economic Perspectives* 9(4), 97–118.

Repetto, R. (1995): *Jobs, Competitiveness and Environmental Regulation: What are the real Issues?* World Resources Institute, Washington DC.

Simpson, R.D.; Bradford, R.L. (1996): Taxing Variable Cost: Environmental regulation as Industrial Policy, *Journal of Environmental Economics and Management,* 30, 282–300.

Sorsa, P. (1994): Competitiveness and Environmental Standards: Some Exploratory Results, Policy Research Working Paper 1249, World Bank, Washington DC.

Sprenger, R-U. (1996): Paper presented on the Effects of Environmental Policies on International Competitiveness, Madingley Hall, Cambridge

Tobey, J.A. (1990): The Effects of Domestic Environmental Policies on Patterns of World Trade: An Empirical Test, *Kyklos,* 43,191–209

US Department of Commerce (1993): *Pollution Abatement Costs and Expenditures, 1991,* Economics and Statistics Administration, Bureau of the Census, Washington, DC.

Wallace, D. (1995): *Environmental Policy and Industrial Innovation, Strategies in Europe, the US and Japan,* Earthscan, London.

4 Environment and Competitiveness of Companies

Thomas Dyllick

In various ways environmental issues become economically and strategically relevant management tasks. To start out, let us look at some illustrative examples. For instance, real estates supposed to be securities for banks turn out to be insecurities, due to contamination of their soils. As a consequence, banks have to find solutions for handling environmental credit risks and for avoiding them in the future. The dyestuff industry is faced with the demand for unpolluting and solvent-free products, since the use of its products causes problems in the automotive and textile industries, some of their major customers. This shows how environmental problems of customers turn into economically relevant problems of their suppliers. Profound changes in the Swiss agricultural policy favour environmentally sound ways of production and, together with changes in demand, lead to a boom for organic food. The Swiss Federal Agency for Energy provides a further type of a new incentive with its publicly effective promotion of energy-saving office equipment in the context of its "Energy 2000-program".

The present text aims at clarifying the complex relationship between environment and competitiveness of companies. In the first place, this relationship needs to be better understood, but then it needs to be effectively managed too.[1] Appropriate theoretical concepts as well as instruments will be developed for this purpose.

Environment and Competitiveness – Understanding the Relationship

Usually, the relationship between environment and competitiveness is represented either as a mere cost problem or as an opportunity for differentiation. Although

---

[1] The present concept for analyzing and managing the relationship between environment and competitiveness is based on the book T. Dyllick/F. Belz/U. Schneidewind (1997). It has been developed in the context of a 4-year research program on "Environment and competitiveness of companies and industrial sectors" at St. Gallen University's "Institut für Wirtschaft und Ökologie" (Institute for Economy and the Environment), Switzerland, under the direction of the author. At present, the concept is elaborated in the context of a subsequent project "From eco-niche to ecological mass market", which focuses on the sector of nutrition. Both projects are part of the Swiss National Science Foundation's "Schwerpunktprogramm Umwelt" (Priority Program Environment).

these perspectives are not wrong, looked at more closely they turn out to be only a part of a much richer picture. In the following, we will analyze more closely this relationship. We will look at different ways how environmental issues become competitively relevant. We will point at the ambivalence of the relationship and its malleability. And we will develop a concept that explains how environmental problems are transformed into competitively relevant problems.

How the Environment Becomes Competitively Relevant

As the introductory examples show, there are different ways of how the environment becomes competitively relevant. In a systematic analysis six such ways may be distinguished:

Political Restrictions

The still common way of environmental policy is that of interdictions and impositions. Many political restrictions are found, above all, in the fields of energy, waste, hazardous substances, risks, and traffic. They are a source of threats as well as opportunities for companies and whole industrial sectors.

Political Promotions

Recently, political programs have increasingly created incentives to support environmentally desirable developments, for example, environmental compensatory payments in agriculture and the promotion of energy-saving office equipment in the context of the program 'Energy 2000'. Political instruments range from subsidies, e.g. to foster public traffic or to develop energy-saving technologies, to the establishing of eco-labels or to creating possibilities for certifying environmental management systems.

Public Pressures

It is due to public pressure that today chloride-free pulp bleaching has largely become the standard in paper production, and coolants in refrigerators are free of CFCs. It is also due to public pressure that food producers renounce the use of genetically modified organisms in cheese production, that asbestos has disappeared from most applications in the construction sector, and that chemical firms today report comprehensively and detailed about their environmental performance.

Avoiding Liabilities and Credit Risks

Environmental risks are assessed by insurance companies as particularly difficult risks. This is shown by the fact that insurance products no longer fully insure environmental liabilities (so, damages due to gradually increasing environmental impairments are excluded from employer's liability) or even completely exclude them from insurance coverage (c.f. waste deposits). Meanwhile, all major swiss banks have established special offices and made special provisions to detect and assess environmental credit risks. These developments show that today pressures as well as incentives to further minimize environmental risks are of a considerable business impact.

Greening the Supply Chain

In their supply activities companies usually have the power to make environmental claims and to enforce them effectively. The power of demand is used both for defensive and for offensive reasons. In order to safeguard themselves against liability risks, firms require information about contents and manufacturing procedures of the products they purchase. Car producers and other 'just-in-time' industries also audit their suppliers with respect to environmental performance and risks so as to protect themselves against deficient or delayed supplies. Environmental auditing of suppliers is part of environmental management systems which are being increasingly used on a global scale.[2] Particularly companies differentiating themselves by means of environmental marketing have to rely on the environmental soundness of all elements included in their products.

Environmental Marketing

Environmental marketing is used in many markets for differentiation purposes. This holds for areas as different as organic food, packaging, organic textiles, recycled paper, biodegradable detergents, or low-sulphur eco-fuel. As soon as such technologies or products are put on a sensitized market by one competitor, environmental considerations become effective by the way of direct competition.

This shows that there are quite different ways for the environment to become competitively relevant. Apart from market mechanisms – be they used by competitors (environmental marketing), by customers (supplier auditing) or by insurance companies and banks (risk assessments) – we find public influences (neighbors, environmental organizations, the media) and political processes (legislation, law enforcement). The environment becomes competitively relevant in a wide variety of ways and in less familiar forms than is found in the context of conventional, strictly market-controlled interactions.

Ambivalence of the Environment-competitiveness Relationship

Companies often underrate or even ignore the strategic relevance of environmental issues. They do so because of the inevitably ambivalent relationship between environment and competitiveness. They often perceive environmental activities as a mere cost factor, leading them to avoid respective measures as long and as far as

---

[2] The Institute for Economy and the Environment is monitoring regularly the Swiss ISO 14001 certifications. They are publicly available under: http://www.iwoe.unisg.ch.

possible. Yet such a perspective ignores the fact that environmental activities are economically interesting if by way of energy management, resource management, waste management, or risk management not only environmental effects but also costs are reduced or efficiency improved. This shows that environmental activities are more than just an undesired cost factor, they can be and they are an economically desired productivity factor as well.

Looking at the strategic level instead of the operative level, again environmental activities appear in a contradictory way. On the one hand we perceive the risks for existing products and technologies, if we think of the ban on substances like CFCs and phosphates or on PVC-products. On the other hand, we perceive opportunities for differentiation, if we think of detergents in concentrated form, chloride-free bleaching of paper, or organic food and cotton.

As a result this ambivalence has to be seen as being Janus-faced on two different levels: on the operative level, it appears as cost factor and as productivity factor, on the strategic level it appears as risk potential and as differentiation potential. Whether the environmental dimension is understood as an opportunity or a danger and whether the strategic potential is exploited or not, therefore depends largely on how a firm or its management perceive this relationship and deal with it in practice.

The crucial question is whether the firm pursues a defensive strategy trying to buffer its present position (left side of Fig. 1) or whether it pursues an offensive strategy aiming at differentiation (right side of Fig. 1). An environmental buffering-strategy is defensive in character. It is aimed at buffering existing advantages, based on technological processes or production sites against environmentally motivated threats. In most cases, these threats are based on public policies or public pressures representing environmental push-factors. Site-related aspects usually are in the forefront here, whereas product-specific aspects are only in the background. Looking at offensive strategies aimed at environmental differentiation the situation is very different. This strategy aims at exploiting existing potentials or developing new potentials for differentiating a company in the market. Market demands are of central importance here, representing environmental pull-factors. Accordingly, products usually are in the forefront here, while site-related opera-

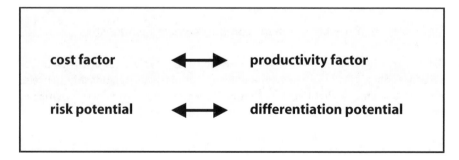

Fig. 1.   Twofold ambivalence of the environment-competitiveness relationship

tional processes are in the background. Hence it is a company's basic orientation and strategy that determines whether the left side or the right side of the twofold ambivalent relationship is of greater practical relevance.

Malleability of the Environment-competitiveness Relationship

The relationship between environmental and economic goals is not as simple as it may seem at first sight. Two opposite positions can be found usually: a conflict model and a complementarity model. The conflict model is based on a contradictory view of economic and environmental goals. It is expressed by the following position: "Environmental activities mean higher costs and lower rates of return". It may be illustrated by the case of the authorities requiring an expensive clean-up of a company's waste or emission situation. Consequently, economically minded managers will engage in environmental protection as little as possible and delay necessary measures as long as possible. The complementarity model is based on a harmonic view of economic and environmental goals. It is experienced in many other situations and can be expressed by a position like the following: "Environmental protection is in the best interest of the company's business". Examples point at reduced costs through environmental activities, for instance by way of reducing waste, reduced risks, for instance by a new storage concept, or improved competitiveness, for instance by a take-back system for used products. Further examples illustrating one position or the other will be easily found.

Since obviously both positions are correct in certain situations, we have to conclude that the simple question regarding the relationship between environmental and economic objectives has no simple answer. An appropriate representation of this relationship can be found in the model of two intersecting sets as shown in Fig. 2. Such a model illustrating the relationship between environment and economy is used by various authors, e.g. Pfriem 1995, p. 93, and Freimann 1996, pp. 356ff, Schneidewind 1998, pp. 48ff.

The intersecting sets model shows how the set of decisions can be divided into different subsets: subsets that are in conflict and subsets that are complementary. In other words, there are decisions that make sense only economically or only environmentally. Yet others make sense in both ways. The size of the intersection, however, cannot be determined logically. It can only be determined empirically, with regard to a specific situation and context. Moreover, the intersection is not static but dynamic. It is affected by stakeholder activities and a changing regulatory and public context as well as by actions taken by the focal company itself. Accordingly, environmental management can be perceived of as management of the intersection. Respective management activities can be located on an operative, a strategic, and an institutonal level. Operative intersection management (e.g. measures for company-internal energy recovery) aims at exploiting the potential of an existing environmental-economic intersection. Strategic intersection management (e.g. the launch of organic food programs by giant Swiss food retail chains 'Coop'

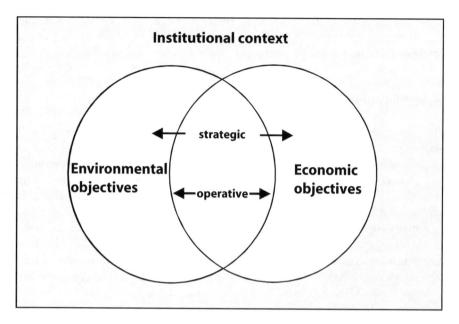

Fig. 2.   Malleability of the relationship between environmental and economic objectives in the context of an intersecting sets model

and 'Migros', or the large scale substitution of waste for fossil fuels in the Swiss cement industry) and changes of the institutional framework (e.g. the introduction of a tax on volatile organic compounds in Switzerland 1997), however, may enlarge the existing intersection.

Environmental Transformation Process

In order to integrate conceptually environment and competitiveness the model of the environmental transformation process may be of help.

The model of the environmental transformation process provides an answer to the question in which ways and how fast environmental problems, once they are known, are able to affect the competitive arena. A general pattern of transformation is of the following form: Initially, there are the environmental problems relevant for a specific industry or an economic sector. They become apparent due to scientific insights that meet with certain public and political interests. Once a scientific insight is in 'resonance' with public or political sensitivities, it then becomes a public issue and, one step later, a topic on the political agenda. If political pressures continue, it will then answered by some form of regulation, whereupon it will influence markets and competition. In view of an environmentally sensitized public and well-organized environmental interest groups pressing for political action,

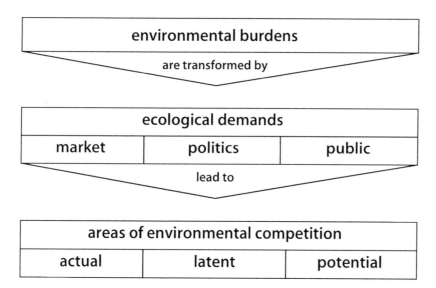

Fig. 3. Environmental transformation process
Source: Dyllick, Belz and Schneidewind (1997)

this process may evolve quite fast. It will slow down, however, when other problems move up on the public and political agenda, as is the case during recessive phases of the economic cycle.

Environmental challenges may be transformed simultaneously or successively by several and different external controlling systems. Market and politics are not the only controlling systems that bring about such transformations, the public plays an important role as well with regard to environmental issues. Moreover, the usually lengthy process of transformation by the official public policy process may be shortcut by organized stakeholder action. If that happens, companies can no longer wait for the signals of change reaching them via the familiar paths of the market. They have to monitor political and public processes as well. The effects evoked by these external controlling systems may alter the competitive arena, forcing the companies and industries concerned to adapt.

A Step-by-Step Approach for Analyzing and Managing the Environment – Competitiveness Relationship

A step-by-step approach for analyzing and managing the relationship between environment and competitiveness will now be developed. It consists of five steps: defining the environmental product life cycle, assessing environmental burdens, assessing environmental demands, defining areas of environmental competition, and defining eco-competitive strategies. We will explain and illustrate each step separately by taking the example of the computer industry.

Defining the Environmental Product Life Cycle

The attempt to assess the product-related environmental burdens poses some real difficulties. Let us, for instance, look at a personal computer (PC) and ask, what environmental burdens it causes. First, we have to think of the energy consumption during its use. When it is used no longer it becomes waste, adding to the total waste output. Apart from highly problematic components like PCB-containing capacitors, which are hazardous waste, it also contains various plastics, which may partly be recovered as raw materials if properly recycled. Focusing on the production, we have to consider heavy pollution of water and soil as well as health and safety risks, due to the use of a wide range of chemical substances. And finally, its manufacturing requires many primary products that come from sectors like the chemical industry, the plastics industry, or the engineering industry.

This example of the PC highlights the interconnectedness of environmental aspects. We are forced to look beyond the limited domain of a single company (here, for example, the Swiss computer dealer that sold the PC), or a single industry (here, for example, the computer industry), and take into consideration preceding stages in the computer life cycle (here, e.g., the plastics industry, the chemical industry, computer manufacturing, transport) and subsequent stages (here, e.g., use, recovery and disposal of the computer). The perspective of the environmental product life cycle reveals that product-specific environmental problems may not only occur within the domain of action and responsibility of one company or one industry, but they may occur at any stage of the usually extended and complicated product life cycle. Nevertheless, problems arising anywhere along the product life cycle will be of consequence for other actors alonf the life-cycle as well. The product life cycle has to be seen as a chain, linking actors of all its stages, from the producers of

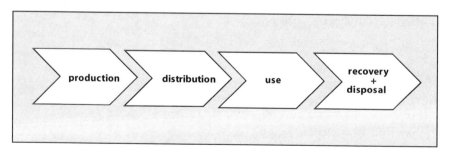

Fig. 4. Stages of the computer life cycle
Source: Paulus (1996), p. 156

raw materials, primary products, or equipment, via the manufacturer of the finished product and the distributor, to the actors in charge of post-consumer waste disposal. The stages of the product life cycle typically span different industries. With regard to a PC, four stages of a computer life cycle can be distinguished: production, distribution, use and recovery/disposal.

Assessing Environmental Burdens

Following to the logic of the environmental transformation process, we start with assessing the environmental burdens on the level of material- and energy-flows. On this level, scientific insights are available, based on objective data and hard facts like, e.g., energy consumption or use of materials. The environmental burden matrix has been developed as an instrument to provide a qualitative, structured overview of the environmental burdens of the whole product life cycle (Fig. 5). Horizontally, the stages of the computer life cycle are listed, vertically we find the relevant types of environmental burdens. Both dimensions, the stages of the product life cycle and the types of environmental burdens, have to be specifically defined for the product or industry assessed. The individual cells of the matrix represent the environmental burdens of the product or industry with respect to the specific stage of the product life cycle and the respective environmental dimension. Only a rough weighting is performed (black = high, grey = medium, white = low), according to the logic of an ABC-analysis, which may be seen as sufficient for strategic purposes.

The environmental burden matrix for personal computers, for instance, reveals that the production stage causes the highest environmental impacts. The major environmental burdens at this stage concern waste, water pollution, and safety/health. Before the computer on its way from the South-East-Asian producer reaches the Swiss dealers and customers, it is usually transported thousands of kilometers

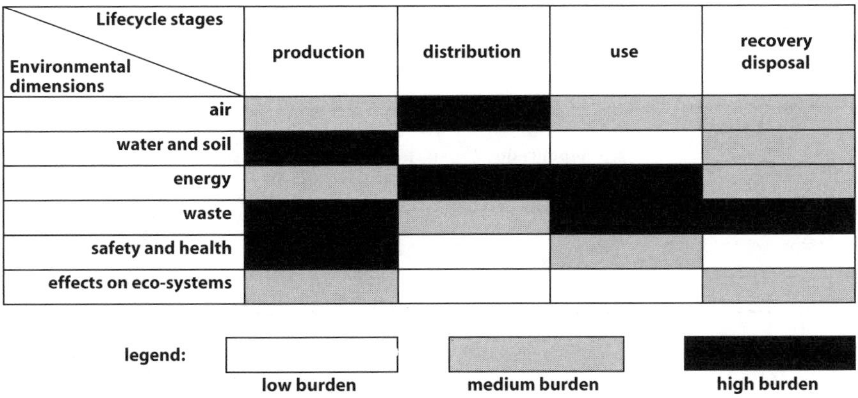

Environmental dimensions \ Lifecycle stages	production	distribution	use	recovery disposal
air	medium	high	medium	medium
water and soil	high	low	low	medium
energy	medium	high	high	medium
waste	high	medium	high	low
safety and health	high	low	medium	low
effects on eco-systems	medium	low	low	low

legend: □ low burden   ▨ medium burden   ■ high burden

Fig. 5.   Environmental burden matrix: personal computers
Source: Paulus (1996), p. 158

by ship and truck, which contributes to energy use and air pollution at the distribution stage. At the use stage of the PC, energy use represents the highest impact. In 1991, the aggregated energy used by computers amounted to more than 5% of the total Swiss electricity consumption. Apart from that, a considerable amount of waste results from used toner cartridges, ink cartridges etc. And contrary to the dreams of the "paperless office in the computer age", the use of paper has doubled since the introduction of PCs. At the fourth and last stage, the continuously increasing amounts of post consumer waste are critical. In 1992 computer waste in Switzerland amounted to 10 000–15 000 tons. For the year 2000, about three or four times this amount is expected. If computers cannot be disposed of in an environmentally sound way, the resulting pollution of water, air and soil will be considerable (Paulus 1996, pp. 155–183).

Assessing Environmental Demands

According to the logic of the environmental transformation process, environmental burdens will turn into environmental demands once picked up by stakeholders. It generally holds that environmental burdens as such are of little direct relevance for companies. They gain importance only if and to the extent that they become incorporated into societal demands, political regulations or changing demands in the markets. So, the public, politics, and the market can be looked upon as representing significant external controlling systems with regard to environmental demands, each exerting control in its specific way. The influence of the public as a controlling system is based on public pressure, the influence of politics on the application of political and governmental power, and the influence of the market on the effects of prices and competition.

In order to assess and analyze environmental demands, we developed the environmental demand matrix (Fig. 6), modelled very similarly to the environmental burden matrix. Horizontally, the stages of the product life cycle are listed, vertically

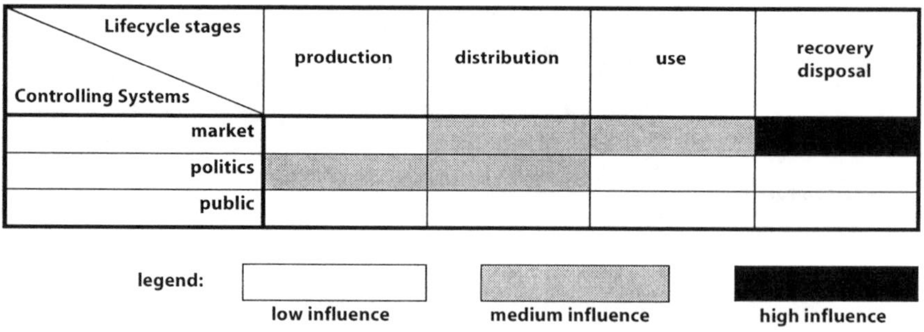

Lifecycle stages Controlling Systems	production	distribution	use	recovery disposal
market		medium	medium	high
politics	medium	medium		
public				

legend: low influence   medium influence   high influence

Fig. 6. Environmental demand matrix: personal computers
Source: Paulus (1996), p. 183

we find the three external controlling systems: market, politics, and public. The individual cells of the matrix show the extent to which environmental demands on the part of an external controlling system exert an influence at the respective stage. As in the case of the environmental burden matrix, we distinguish only three degrees of influence (black = high influence, grey = medium influence, white = low influence).

With regard to personal computers, we see that environmental demands originate mainly from market and politics (Paulus 1996, pp. 183–212). The public still looks upon this industry as being clean. For professional users energy consumption of the computers during use is of relevance. Further important criteria are CFC-free production and environment-friendly recycling/disposal of the computers. But it cannot be overlooked that environmental criteria, generally, are of little importance compared with economic and technological criteria. This is particularly true for the 'home/small office' segment.

Comparing the results of the environmental demand matrix with those of the environmental burden matrix, we are able to assess the "environmental appropriateness" of demands addressed at computers. Looking at the later stages use and recovery/disposal, environmental demands match quite well with environmental burdens. The production stage, however, with production taking place in other parts of the globe and being accompanied by high environmental burdens, is not evaluated appropriately by environmental stakeholders and external controlling systems. The same holds true for transportation, where a discrepancy exists, which holds the potential for future demands.

Defining Areas of Environmental Competition

Areas of environmental competition exist where environmental problems may lead to competitive advantages or disadvantages depending on whether they are solved or not. Both, costs and possibilities for differentiation are important to look at in this respect. Relative cost advantages exist when companies succeed in keeping costs for environment protection lower than competitors, subject to the same or similar conditions. In most cases, cost advantages are of this relative kind, whereas absolute cost savings through environmental measures are more difficult to achieve. Advantages through differentiation result from modifying existing products by adding an environmentally useful feature (e.g. color pigments without heavy metals) or when new products or services are developed in reaction to environmentally induced business opportunities (e.g. remediation of contaminated soils etc.).

Areas of environmental competition can be classified according to their state of development as actual, latent, or potential. In actual areas of environmental competition environmental strategies have become a major aspect of the competitive dynamics. An example in the Swiss case is organic food, where bio-food has reached a 10–30% national market share in certain product ranges (milk, vegetables). In latent areas of environmental competition environmental strategies have become

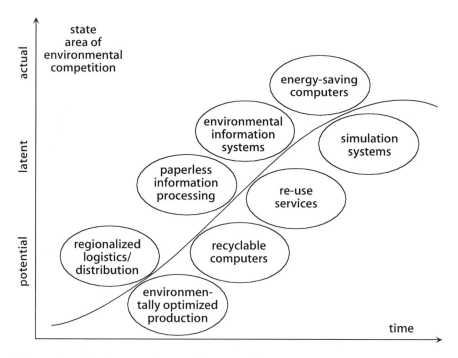

Fig. 7.   Areas of environmental competition: personal computers
Source: Dyllick, Belz and Schneidewind (1997)

a minor aspect of the competitive dynamics, typically being restricted to market niches. (e.g. less than 5% market share) This is typical for the early phases of environmental products or services. Often they are launched by small, pioneering companies or they are restricted to certain segments (e.g. high-end products) or regions (e.g. food from regional producers). In potential areas of environmental competition, the transformation of environmental problems into solutions offered in the market is still in an early stage of development. In this case environmental problems have been recognized and technologically as well as economically feasible solutions are being explored, without having resulted in acceptable solutions yet. Fuel cells for cars may be an example here.

As to the areas of environmental competition in the computer industry, they are located between two different trends: "ecologizing the computer" on the one side, "computerizing environmental management" on the other side. While the first trend relates to the environmental problems caused by the computer itself, the second trend points to environmental problems which can be (better) solved by using computers (Paulus 1996, pp. 369–387).

With respect to ecologizing the computer, only energy-saving computers can be looked upon as an actual area of environmental competition today. The decisive triggers are the consumers' interest in the costs of PC-use and energy-consciousness in buying on the part of public bodies, e.g. the U.S. government. The other actual areas of environmental competition are part of the second trend, computerization of environmental management, for example environmental information and simulation systems (e.g. in the development of vehicles or pharmaceuticals) and technologies allowing paperless information processing.

Due to the environmental emergency situation in the field of computer disposal, the areas of recyclable computers and re-use services are in a latent state. Many producers have developed and offered solutions, but they have not yet gained much interest. Approaches for ecologizing production processes and reducing transport volumes in computer production and distribution, on the other hand, are of no competitive relevance so far, even though they are related to significant environmental burdens. They will not move up to become a latent or even actual area of environmental competition, as long as technological imperatives (e.g. ever more differentiated chip technologies) and the exploitation of global economies of scale remain predominant.

Defining Eco-competitive Strategies

In the environmental context it is not just the market that plays a role but it is society also, if we think of politics and the public. Empirical studies about the importance and influence of stakeholders in the environmental context show that managers tend to perceive society to be even more influential than the market. (Dyllick 1996, pp. 9ff, Baumast/Dyllick 1998, pp. 24ff) Therefore, apart from the market orientation of eco-competitive strategies, we always have to consider its societal orientation as well. Furthermore, we can distinguish between a defensive and an offensive strategic behavior. From these two dimensions we can derive four types of eco-competitive strategies.

The environmental market buffering strategy is based on defensive behavior and is oriented towards societal influences that threaten existing markets and businesses. Before public and political demands become competitively relevant, companies react to them out of self-interest. They try to influence the environmental transformation process by communication strategies or by accepting and following self-imposed restraints. For instance, in the context of the "Responsible Care" program, chemical companies commit themselves to a comprehensive environmental and safety management system in order to minimize existing risks and to improve their public image. This strategy aims at being perceived as environmentally "clean" and at buffering existing markets or businesses against restraining environmental demands.

Like environmental strategies to buffer existing markets, environmental cost strategies are based on defensive behavior, but they are oriented towards the market. Companies do not intervene actively into the environmental transformation process, rather they take its outcomes as given and try to meet environmental demands as "efficient" as possible, that is, as cost-effective as possible. Often this leads

strategic behavior \ strategic orientation	society	market
defensive	environmental market buffering strategies ("clean")	environmental cost strategies ("efficient")
offensive	environmental market development strategies ("progressive")	environmental differ- entiation strategies ("innovative")

Fig. 8. Typology of eco-competitive strategies
Source: Dyllick, Belz and Schneidewind (1997)

to solutions that reduce total costs, too. Environmental management, similar to business reengineering, looks at business activities from a new perspective. Thereby, it circumvents conventional patterns of perception and action, very often leading to more efficient and cost-saving solutions. One such perception barrier is, for instance, the narrow definition of environmental costs, taken as direct costs of waste disposal or reduction of emissions. If applying "environmental total cost accounting" instead, indirect costs associated with producing waste and emissions as well as with purchasing the materials, that will be transformed into waste and emissions later, are taken into consideration as well. This allows to perceive new environmentally and economically advantageous solutions. A second perception barrier extends the view so as to include costs that are incurred at stages preceding or following a company's activities. Knowing the structure and location of the complete "environmental life cycle costs" offers new directions for reducing costs outside the own company. This may be very interesting from a strategic perspective, offering possibilities for differentiation in the market.

Environmental differentiation strategies are based on offensive behavior and are oriented towards the market. Their objective is to use environmental strategies for creating innovation potentials, i.e. for being "innovative" in the market. Especially in saturated markets, in which products become more and more indistinguishable and exchangeable from a consumer's view, environmental strategies have shown to be an interesting dimension for differentiation. But possibilities for a company to differentiate itself by environmental products or services can be found also in newly emerging environmental markets. In contrast to conventional differentiation strategies, environmental differentiation strategies rely heavily on qualities such as a company's public credibility.

Eco-products and eco-markets often face the danger of remaining caught in niches. For instance, this has been the case with solar collectors. Though environmentally sensible and technologically feasible, solar collectors have not been able to win due recognition in the market on a large scale so far. Such a development would need changes in the institutional framework. To promote such changes environmental market development strategies are pursued. These strategies have an offensive character and are oriented towards the societal field influencing and defining the scope and rules of competition in the market. They aim at shaping these preconditions so as to create and enlarge areas of environmental competition. Strategies of this type are called "progressive". They focus on the environmental transformation process and try to promote and accelerate it by taking appropriate measures. Examples of this strategy are the efforts of the Swiss food retail chains "Migros" and "Coop" to develop and provide wide access to eco-balance data for packaging material, or of AEG household appliances in Germany to support the introduction of an environmental tax reform.

Typically, but not necessarily, these four strategies occur in a sequence, starting with market buffering strategies, continuing with environmental cost and differentiation strategies, until market development strategies will be reached. This sequence is due, above all, to organizational learning processes and developments that cannot simply be skipped. The first two strategies are pursued most often. Differentiation strategies occur more rarely, and only a few pioneering firms employ also strategies to develop the market. In the computer sector, environmental market buffering strategies and cost strategies prevail. Individual attempts to follow a more innovative differentiation strategy (e.g. IBM's efforts for an eco-PC) have not been notably successful so far in this industry. Strategies in environmentally more advanced industries, e.g. food, textiles, packaging, paper etc. have been more successful in this respect.

References

Baumast, A.; Dyllick, T. (1998): Umweltmanagement-Barometer Schweiz 1997/98. IWÖ-Diskussionsbeitrag Nr. 59, St. Gallen.

Dyllick, T. (1996): Umweltmanagement-Barometer Schweiz 1995/96. Überblick über die Ergebnisse, IWÖ-Diskussionsbeitrag Nr. 37, St. Gallen.

Dyllick, T.; Belz, F.; Schneidewind, U. (1997): Ökologie und Wettbewerbsfähigkeit von Unternehmen. Carl Hanser and NZZ Buchverlag: Munich and Zurich.

Freimann, J. (1996): Betriebliche Umweltpolitik. Paul Haupt (UTB): Bern.

Paulus, J. (1996): Ökologie und Wettbewerbsfähigkeit in der Computerindustrie. Ph.D.- Thesis, University of St. Gallen, Switzerland. Difo Druck: Bamberg.

Pfriem, R. (1995): Unternehmenspolitik in sozialökologischen Perspektiven. Metropolis: Marburg.

Schneidewind, U. (1998): Die Unternehmung als strukturpolitischer Akteur. Metropolis: Marburg.

5 From Environmental Management Towards Sustainable Entrepreneurship

Hugi H. Kuijjer

In The Netherlands we now face the challenge of reconciling two objectives at the same time: greater prosperity and protection of the environment. In this article I want to share with you some Dutch experiences on how we try to overcome this dilemma of a drastic improvement of environmental performance of industry, while at the same time strengthening its competitiveness. Or to realise a sustainable society within the span of one generation (carrying out the National Environmental Policy Plan) and an annual 3% growth of the Dutch economy.

The key for achieving this is consensus and sharing responsibilities between industry and government. 'The Dutch Green Polder Model'.

Introduction

The cornerstone of Dutch environmental policy is the National Environmental Policy Plan (NEPP). This aims to tackle all the existing environmental problems and to reach a sustainable society within a single generation. Achieving this aim will require a considerable amount of effort. The NEPP lays down clear objectives in respect of all forms of environmental impact. These objectives are very ambitious ones. In order to ensure a stable quality of the environment, the emission reduction targets to be met by industry are 50–70% for the year 2000 and 80–90% for the year 2010. This will certainly not be achieved with the conventional instruments for managing environmental impact. The traditional 'end-of-pipe' measures are simply inadequate to achieve these emission reductions without standing in the way of a doubling of the GNP by the year 2010. Calculations show that trying to manage environmental impact via 'end-of-pipe' measures ultimately becomes not affordable.

So the execution of the NEPP requires far-reaching changes in our production and consumption processes and radical technological, social and economic changes are necessary. On the environmental agenda three basic techniques lie at the heart of these changes: integrated chain management, extensive use of energy and raw materials, and quality improvement.

As a result of the ambitious objectives environmental policy has to be more integrated into economic decision making. It is clear that Government can never, on its own, bring about the necessary change to a sustainable society. The required changes are only feasible if there is sufficient support and if it becomes clear what long-term effort sectors of industry and of course individual companies will have to make in order to attain the necessary improvement of the environmental quality. So we also have to make a major step from a command and control approach towards an approach based on co-operation between government and industry.

Phases in Governmental Policies

To get a good understanding of experiences in The Netherlands, I first want to give a brief overview of the different phases in the development of policies towards sustainable industrial production from both sides: government and business. Roughly spoken these different phases can be recognised all over the world. The figure shows indirect or direct governmental control on industrial environmental performance (vertical axis), and a distant/confrontational or a close/co-operative relationship between business and government (horizontal axis).

If the development of environmental policy is analysed it is almost sure that these different phases, or in other words, approaches will be recognised.

First there is a situation where business is largely 'unconstrained' (1) by environmental requirements of any sort. Due to public awareness and political pres-

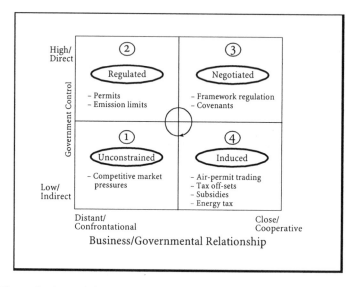

Fig. 1.   Phases development in governmental policies

sure, strict environmental controls are put on companies, mainly by unilateral action taken by government.

In the 'regulated' phase (2) legal instruments like permits and discharge limits are developed, often related to single media like air pollution, waste or noise. In many cases legal requirement is very detailed and enforcement bodies are built up. Command and control approaches can provide important environmental improvements and they also send clear signals to business that changes have to be made. However, in the long term command and control approaches are not effective. It does not stimulate industry to find 'big step' solutions. Rather it encourages 'end of pipe' measures and 'small step' improvement. And when the environmental performance is already quite high, further steps to reach sustainable industrial development become very expensive.

After picking the 'low hanging fruit', environmental policies have to become more integrated in nature and have to have more real impact on companies. This calls for another approach which needs more flexibility and which appeals to the own responsibility of industry to deliver continuous environmental improvement. This in industries own interest and in relationship to the continuity of companies. These type of approaches can be found in the diagram under phase 3 and 4.

In the 'negotiated' phase (3) the government sets targets and objectives, but also makes provisions for negotiations between authorities and industry. These negotiations allow industry to influence the targets and objectives, and to set a suitable time scale. Industry is largely left free to determine the means by which the targets and objectives will be met. Working on the same, agreed environmental agenda this approach can be characterised as close co-operation between government and industry. The covenant approach in The Netherlands is an example of this.

The 'induced' approach (4) includes policy instruments such as air permit trading, tax off-sets, govern-mental subsidies and energy taxes. These instruments encourage, rather than force industry to improve further on. They reward companies which invest and make improvement possible which go beyond day to day standards.

An ideal situation is to achieve again the unconstrained phase, in which government control is replaced by a competitive market. This happens when competitive forces in the market impose all the required incentives for sustainable entrepreneurship and companies link their integrated environmental policy with their financial-economic policy.

I briefly described the different approaches predominantly made in Governmental policy in general. In The Netherlands after a more or less unconstrained phase till the beginning of the seventies and a command and control approach during the mid seventies and the eighties, we made an important step on the road to sustainable development with the publication of the NEPP at the end of the eighties. The NEPP has to be carried out in close co-operation with relevant groups in Dutch society. So with the implementation of the NEPP we are now in the negotiated phase. The way we do this in the most relevant sectors of industry is called 'Target-group policy for industry'.

Target-group Policy for Industry

The Target-group policy sets out by making clear what is expected of the target-group in question. For example, the statement "The emission of benzene should have been reduced by 75% by the year 2000" does not mean a thing, if it is not indicated which sources must be responsible for this reduction. The execution of the NEPP for industry, therefore, started with the translation of NEPP objectives into specific emission-reduction objectives for industry as a whole. These in turn were translated into the various sectors of industry. The next translation step was the one from the sector level to individual companies. Environmental policy for industry in The Netherlands is largely carried out by local authorities which issue permits to companies. In doing so licensing and enforcement procedures must be fine-tuned to the realisation of NEPP objectives. This has already been realised by the Target-group policy for industry for several sectors, such as: Primary metals, Chemical, Printing, Dairy, Metal and electrical, Textile and carpet, Paper and cardboard, Abattoirs and meat processing.

The approach focuses on the drawing up of a declaration of intent (a so called 'covenant') in which industry and authorities agree that the policy of the parties involved be aimed at the realisation of an Integral Environmental Objective (abbreviated IEO) for the sector in question. This objective constitutes a summary of the NEPP objectives for the sector. The agreement means that industry itself will take the initiative to adopt the required environmental measures. For authorities, the agreement signifies that the development of their policy and the licensing and enforcement tools will be attuned to the initiatives of the companies in question. The covenant also lays down agreements on the way in which the IEO for the sector will be worked out for individual companies. The elaboration depends on the nature of the sector. Large and more complex companies are able to draw up their own 'Company Environmental Plan' (abbreviated CEP). In such a plan the company indicates what contribution will be made to the realisation of the IEO for the sector. Smaller companies draw up their CEP by filling in a kind of checklist and picking measures from a given package.

The CEP is drawn up every four years in close co-operation with the local authorities responsible for the licensing of the company in question. If the licensing authorities are able to agree on a CEP this will be the basis for the licensing and enforcement towards the company for the next few years.

Every four years all CEP's are monitored against the sector targets. The bottle necks are evaluated in a Consultative Group at sector level with participation of all involved parties. The results of the evaluation give guidance to the next cycle of CEP's. In this manner, direction is given to the further development of the environmental policy within the industrial sector concerned, including a focus on the correct priorities.

The fundamental principle underlying this approach is that the responsibility for achieving these objectives lies primarily within the sector of industry. Participation in the Target-group policy for industry offers the business community various advantages. It is now easier for companies to set priorities for the environmental problems they want to tackle. In addition, companies are now free to select the most cost-effective environmental measures. Once again, they are able to implement these measures at the moment that suits them best, e.g. when an installation has been shut down for maintenance.

In the past, different authorities came up with different environmental requirements at different times. Within the context of the target-group policy, however, authorities harmonise their activities with each other. In addition the target-group policy makes it clear what the long-term aims of the government actually are. This means that the company can take these into account if, in the interim, they make replacement investments, for example, or undertake new building. This avoids disinvestments.

The government also benefits from the Target-group policy for industry, since it leads to a degree of reduction in environmental impact that could never have been achieved by the traditional method. The target-group policy tackles environmental problems integrally, in their interrelationship. This avoids 'end-of-pipe' measures that often simply postpone environmental problems. The policy's integral approach enables it to find structural, process-integrated solutions that prevent environmental problems at source. In addition, the target-group policy 'filters out' the environmental problem areas. The government can then focus its policy on precisely those areas.

The first results of the target-group approach are very encouraging. Evaluation of the first round of CEP's (Primary metals, Chemical, Printing, and Dairy) show that most of the sector targets for the year 2000 are within reach or have already been met. Of course there are also specific problems to focus attention on such as climate change, and acidification. The mentioned Consultative Group at sector level has completed a detailed analysis of the components for which targets will not be realised in the year 2000. The Consultative Group has come up with proposals how to overcome these bottlenecks. The results of this assessment gives guidance to the next cycle of CEP's which started some time ago. In the most relevant sectors of industry we are now on our way to sustainable industrial development.

Environmental Management

With the Target-group Policy for Industry and the CEP's, which are drawn up by companies, it be-comes clear what (long-term) effort companies have to make in order to attain the necessary improvement of the environmental quality. For an adequate execution of environmental measures, however, it is also required that companies have at their disposal a number of organisational, administrative and technical facilities. An Environmental Management System (or EMS) offers the frame-work for this. One of the conclusions of the NEPP is that environmental management is a necessary tool to realise the NEPP objectives and to attain stable, sound environmental control. In short, an EMS is the tool for companies to realise 'sustainable industrial development'. It is an essential instrument as regards the continuity of a company.

The Dutch Government set down its objectives on the significance and development of environmental management in a memorandum in August 1989.

- In this memorandum the Government set the objective that nearly all 10 000 companies which cause mode-rate to serious environmental impact or specific environmental risks should have an environmental management system in place by 1995. These systems should be geared to the nature, size and complexity of the individual companies.
- In addition, the Government expects the relevant sectors of industry to have taken definite steps towards introducing adequate environmental management in the some 250 000 companies with limited environmental impact.

This policy objective is more of a target to be aimed for than a cast-iron requirement. Of course it does not really matter whether, as of 31 December 1995, it is exactly 10 000 companies that have implemented an EMS. It is more important that the business community makes sufficient progress in the implementation of such systems.

Because EMS' are in trade and industry's own interests the policy of the Dutch Government is that the implementation should be reached on a voluntary basis. That is why environmental management will not be enforced by legislation. Instead the implementation of EMS' was facilitated over a five year period (1990–1995) by an incentive programme with a budget of 50 million Dutch guilders.

The incentive programme made a distinction between different types of project. Projects in specific sectors of industry, for example, were intended to develop a model EMS for an entire sector of industry. This saves each company having to separately devise an EMS. Individual projects meanwhile were intended for further investigation, e.g. researching the possibilities for achieving a more effective harmonisation of licensing procedures with environmental management. In addition, an initial subsidy had also been made via the incentive programme for the establishment of so-called Industrial Environment Services. Industrial Environment Services that were starting up could obtain a subsidy to cover part of their operating costs for the first three years. Industrial Environment Services are non-profit-making institutions that work on behalf of, and are set up and run by the regional business community. They support implementation, advise on environmental matters or support licensing for SME's. Companies have to pay for their membership.

Although the implementation of environmental management should be reached on a voluntary basis it is our policy to periodically investigate how much progress the so-called '10 000 Group' has made in implementing EMS'. Government expects this group, above all others, to implement an EMS.

In 1996 the progress made in implementing EMS' was measured for the third time. This 'Evaluation of Company Environmental Management Systems 1996' again showed evidence of growth. 34% of the companies ('the front-runners') have already implemented an EMS or have almost done so. 37% of the companies ('the candidates') are hard at work implementing such a system. 23% ('the beginners') are introducing environmental measures and have, indeed, al-ready implemented certain elements of an EMS, but are not yet working on the implementation of an EMS. Only 6% of the companies ('the non-actives') are still not doing anything very much about environmental management.

Evaluation '96 showed that more and more Dutch companies are taking environmental management on board. The percentage of non-actives and beginners is falling. Since 1993 a relatively large number of companies have begun developing an EMS. At the same time, the proportion of candidates and front-runners is increasing. Many companies have developed from beginners or non-actives into candidates or front-runners. Companies are, in other words, continuing to strive for improvement in their environmental management. To date, 71% of Dutch companies in the 10 000 Group have lready introduced an EMS or are hard at work developing one. 90% of them want to improve their EMS even further in the future.

Large companies are much more likely than smaller ones to have an EMS. According to Evaluation '96, 29% of companies with fewer than 100 employees are front-runners. Among companies with 100 to 500 employees, on the other hand, that figure is 61%, while 86% of companies with more than 500 employees are in the front-runners group. Although it is a good sign that the larger companies are lea-ding the way, since they are, after all, responsible for by far the greatest amount of pollution in The Netherlands, it is not very good news that small companies are

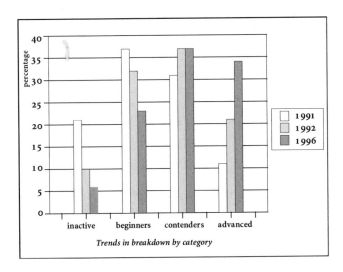

Fig. 2.   Progress of environmental management

lagging behind. Such companies do, after all, make up the largest part of the Dutch business community: some 83% of the companies in the 10 000 Group have fewer than 100 employees.

The percentage of front-runners also differs from one sector to another. The chemical industry, for example, is a clear leader, with 65% in the front-runners group. The foodstuffs industry is not far behind, on 52%. The metal electrical engineering industry is on 36%, while the construction industry lags somewhat, with 21%. Among utility companies/government enterprises, 50% are in the front-runners group.

All in all, Evaluation '96 shows that environmental management is now firmly established in Dutch business community. In fact, it is impossible to conceive of Dutch business life without it. Some of the results suggest that a further growth in environmental management is likely. The overwhelming majority of companies surveyed said that they had actually benefited from the implementation of an EMS. 50% think that implementing an EMS leads to financial savings, among other things through energy savings, waste prevention and savings on raw materials, while 74% think that an EMS reduces environmental risks and 79% think that implementing an EMS has improved their image.

Co-operation Between Industry and Government

Sustainable industrial development will pose new requirements for the environmental management of companies. The heart of the approach should be the actual integration of environmental policy into the broader corporate policy, for example on investments. And so the development of EMS' aimed at continuous improvement of environmental performance, is a crucial factor for success.

At the same time governmental intervention should take greater account of the phase of environmental management the company has reached. In this 'stragglers, people in the middle and leaders' can be recognised. Of course companies do differ in their stance on environmental affairs. Their stance can be 'defensive, following, active, or pro-active'. The position of the local authority should be aligned with the company's stance on the environment and the phase of development of EMS it is in. From the perspective of a company and the local authority four phases can be identified (Fig. 3).

What are the characteristics of the different phases?

1. The 'defensive' company attitude can be described as no or insufficient compliance with environmental regulations, little or no organisational provisions for the environment and no policy or objectives. The general attitude of the company is 'the environment is a burden and a negative operational issue'. Government's stance will be focused on a command and control approach. Within the target-group approach these companies are potential so called 'free riders'.

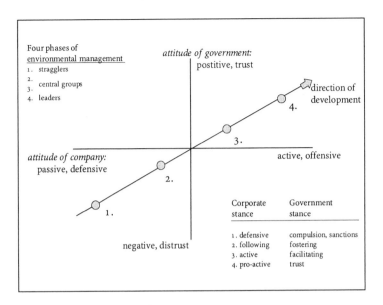

Fig. 3. Development of environmental management

2. A 'following' company is one in which compliance with regulations is the primary objective. If necessary the company will introduce organisational measures. The environment is accepted as a regular issue in business operations. Companies in this phase should be fostered by government, but at the same time it is wise to keep the pressure on.

3. An 'active' company sets itself objectives and targets beyond compliance, while EMS' usually are well developed. In this phase an EMS in accordance with ISO 14001 is typical. Within the company the environment is at the border line between being a regular issue and a real challenge. Government's stance should be facilitating by giving flexibility and encouraging the company to a more proactive attitude.

4. In a 'pro-active' company there is a high awareness of the environmental issue. The environment is seen as a strategic challenge, and is also related to products and natural resources. The relationship with government is based on trust.

It is essential that licensing authorities come to recognise these development phases in practice and attune their way of intervention accordingly. The model also makes clear that a sufficient level of environmental awareness within industries is a pre-condition for progress towards sustainable industrial development. A point to note is that the majority of companies in The Netherlands is now between phase 2 and 3. It must of course be kept in mind that this is only a diagrammatic representation of what in practice is much more complicated.

What we try to achieve is a relationship build on trust between active companies and local authorities. But trust is not something you get for free; you have to earn it and both parties should work on it.

In our view the development of EMS' is not an isolated issue. It is strongly related to other activities that have been deployed within the policy of Dutch Government towards industry. So how do we give shape to the development towards a sustainable industrial development and a different relationship between industry and local authorities in practice? Our strategy is based on:

1. the implementation of adequate EMS' (in accordance with ISO 14001 / EMAS-regulation);
2. the drawing up of company environmental plans (CEP's) every four years;
3. issuing a licence on outlines and enforcement aimed at the adequate operation of the EMS for production sites of (pro-)active companies based on the CEP. This allows industry flexibility in the means to meet the objectives on targets mentioned in the CEP;
4. companies giving account of their activities to authorities and public by a yearly site report on environmental performance. Transparency of environmental management implemented in companies is important in this respect and reliability of company information is a precondition.

In The Netherlands we are working simultaneously on these four activities in close co-operation between government and industry. In this the relationship between the Target-group policy for industry and environmental management is very important. As an EMS is an indispensable tool in setting up and implementing a

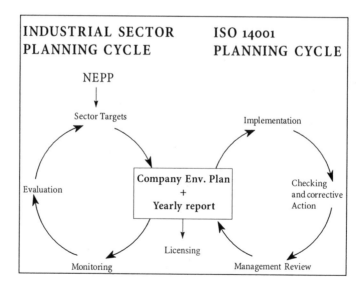

Fig. 4. Planning cycles for the company environmental plan

CEP the covenants always include an agreement stating the date by which participating companies must have introduced an EMS.

According to the ISO 14001 standard an EMS consists of a cycle of 'plan, do, check, and act'. The Target-group policy for industry consists of a comparable cycle. First an Integral Environmental Objective (IEO) is drawn up. Then the companies draft their CEP. These are evaluated, after which the conclusions are incorporated into the following cycle. What Dutch policy is in fact striving for is for these two cycles to be brought together, with the CEP as the connecting link.

Standardisation and Certification

An EMS can be proven to be working well by setting up management systems in accordance with an accepted standard and certification based on that standard. Standardisation and certification of EMS' are being developed in The Netherlands by implementing the EMAS regulation. EMAS stands for an EC directive on the voluntary participation of companies from the industrial sector in a community environmental management and audit system. If companies satisfy certain requirements, they are entitled to EMAS registration and are allowed to hold a declaration of participation. It is in fact a European certificate for an EMS and for the way in which companies report on their environmental performance.

In The Netherlands, the standard ISO 14001 is regarded as the standard for EMS'. Certification on the basis of ISO 14001 and verification on the basis of the EMAS regulation are developed within a single structure. There is only one scheme for certification, one Central Council of Experts and an EMS Certification Co-ordination Office has been set up. This office is also the competent body in the context of the EMAS regulation.

Firstly this all means that the quality of a certificate is guaranteed. And secondly that a company which has a certificate on the basis of ISO 14001 automatically complies with the EMAS regulation, as far as the EMS is concerned. In addition EMAS requires an environmental report, setting forth the company's environmental performance. But the verification of this report is a rather small additional step.

Fig. 5.   Standardisation and certification

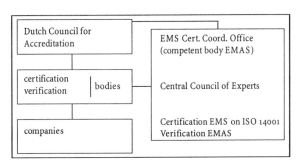

If a certificate guarantees the quality of an EMS it can come to play a major role in the relationship between authorities and companies.

Licence on Outlines

The idea behind Dutch environmental policy is that companies have their own responsibility towards the environment to which government action should be tailored. Reflection of this 'own responsibility' is: to sign agreements on implementing the NEPP objectives, to draw up a company environmental plan (CEP), to implement an EMS' (ISO 14001 / EMAS) and to issue a yearly environmental report.

The government's role is to base its licensing and enforcement on this and to reward progressive companies for their environmental efforts. Companies with an agreed on CEP, an adequate EMS, and which compile a yearly report can obtain, as countervalue, a so-called 'outline licence'. Such an outline licence is more efficient for both the company and the government.

A traditional environmental licence in which the government specifies the precise environmental measures that the company must take is, as a result, less suitable for these progressive companies. Within the command and control approach, the licence issued to an industrial site usually is very extensive. It normally is targeted in detail at the subsequent different primary processes within the site. A traditional licence for this reason contains some hundreds of requirements. The licensing authority prescribes in detail the means to be used by the company to reduce its environmental impact. The reaction of the company on such a detailed licence often is defensive. The enforcement usually can be characterised as unstructured. There is no distinction between main points and side-issues, while enforcement does not tackle structural causes of non-compliance by the company.

The next figure shows the outline licence in diagram form.

Rather than a licence which lays down detailed measures for the various parts of the site, the outline licence contains the objectives for the site as a whole and focuses more closely on essentials. The objectives are derived from the CEP, which is discussed with and approved by the licensing authorities. The EMS translates these objectives into measures for the different primary processes within the site. The company is free to choose the most efficient way of achieving its targets and the most cost-effective measures. Such a licence should also contain regulations on the yearly reporting of the environmental performance, and regulations on notifications of unusual events.

In terms of volume, such an outline licence contains no more than a few pages of paper with a few dozen requirements at the most. In an outline licence as much as possible so-called 'goal-orientated' directives are used. And in this situation, enforcement is focused on main issues and is based on the assumption that the company's EMS has lead to self control.

Fig. 6. Licence on outlines

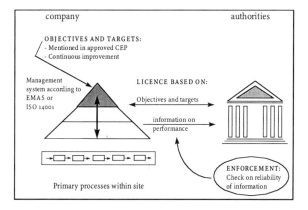

Such a system of licensing and enforcement offers the company much more flexibility in shaping environmental management. Significant savings of environmental costs in the order of 20 to 50% are possible without any compromise to the desired environmental results. In other words, in the past, environmental policies in The Netherlands were considered by industry to be a negative investment factor. Nowadays it is recognised as a positive factor, because of its consistency and its integration in economic decision making. In the meantime major improvements in the environmental performance of industry have been be realised.

Sustainable Entrepreneurship

When industry is on its way towards a more sustainable industrial development, there are three distinct pathways to higher performance: 'incremental, redesign and rethink'.

1. The incremental option is closest to today's methods of environmental improvements. It is based on 'end of pipe' solutions and often related to command and control approaches. This option delivers significant progress in the early stage of development by picking the low hanging fruit. However, on the long run for further improvement, 'end of pipe' solutions become less and less effective and also quite expensive.
2. The redesign option requires the reformulating of products and processes to deliver a major reduction in environmental impact. Redesign options are also known as 'prevention of pollution' or 'process integrated' solutions. Implementing this type of measures is often combined with new investment. It requires investments in research and development and it takes time to achieve. An example is the wide-spread redesign of CFC-using processes.

Fig. 7. Phases development in business policies

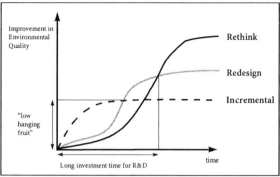

3. The rethink options are the most radical in nature. They require a major shift in patterns of production and consumption, and deliver quantum leaps in performance improvements. The environmental benefits of rethink options are often unpredictable at an early stage and difficult to measure and to manage. However, the implementation of these options within industry and society can accelerate once these new systems and technologies are in place. Examples of a rethink approach are the application of ICT and future solutions of the CO_2 problem.

Figure 7 shows the three pathways to higher performance or in other words the phases in the development of business policies.

As mentioned our aim is to overcome the dilemma of a drastic improvement of environmental performance of industry, while at the same time strengthening its competitiveness. Or to realise a sustainable society within the span of one generation (carrying out the NEPP) and an annual 3% growth of the Dutch economy. This requires far-reaching changes in our production and consumption processes and so radical technological, social and economic changes are necessary.

A shift of emphasis from picking the low hanging fruit by end of pipe solutions and trying to make existing systems more efficient to process-integrated technologies and environmentally-oriented product development is needed. In other words in the coming decade we need to combine continuous improvement in 'small steps' and a transition to fundamental new production and consumption processes by breakthrough technologies in real 'big step' solutions by using all three options: incremental, redesign and rethink. In doing so an absolute decoupling between economic growth and its environmental burden has to be achieved.

'Decoupling' refers to improving living standards (economic growth) while at the same time reducing the environmental pressure. Absolute decoupling occurs when the environmental pressure reduces or at least remains constant while economic activities are increasing: i.e. economic growth while the environmental pressure is falling.

So industry will face new problems, both economic and environmental, in the coming decade. In order to adapt to trends such as globalisation, increasing scale, new technological developments and new standards of quality, industry and services must invest in know-how and quality. Industry and services will become increasingly interwoven with another. On the environmental side the key themes in the medium and the long term will be: dematerialisation (using less materials), increasing energy and material efficiency, closing material cycles, the supply of cleaner products and services and a move from 'end of pipe' to 'process-integrated and product development'. There will also be increasing attention paid to the demand side and in particular the role of the consumer.

As a result of this environmental policy has to be more integrated into economic decision making. It all asks for an approach in which companies link their integrated environmental policy with their financial-economic policy. The first changes towards this perspective can now be noticed, with the environment becoming interesting in commercial, market and image terms. Really leading companies do not see the environment as a cost factor, and as an area requiring more or less isolated attention in response to pressure from the authorities or pressure-groups. They see that society is increasingly demanding that industry adopts a clear position on the environment. And that environmental policy does not stop at the company's gate, but is pursued in relation to the entire life-cycle, 'from cradle to grave'. This is leading to new types of relationships between companies (including the financial institutions) in which they rate one another increasingly in terms of environmental performance and risks. They are convinced that these changes are not a threat to industry but, on the contrary, will create new market opportunities and will lead to continuity. The environment is becoming an opportunity!

As the environment gains importance in the marketplace, the environment is assuming a larger role in the choice of processes and products, and has to be more closely integrated into the general commercial strategies and thus is becoming a management matter of the board.

This new role for the environment necessitates new management instruments for the board. So one of the spearhead programmes of the Dutch policy document on Environment and Economy (1996) is to develop 'Sustainable entrepreneurship' (or strategic environmental management) as such a new tool.

Sustainable entrepreneurship involves an approach in which a company links its integrated environmental policy with its financial-economic policy. The purpose is to build synergy between environmental and economic aspects of running the company and to transform the costs of environmental measures into advantages such as improved market position and more effective realisation of environmental objectives. Sustainable entrepreneurship therefore seeks for closer links between existing environmental instruments at the strategic level and to relate environmental performance to economic performance. Eco-efficiency appears to be a concept embodying the principles of sustainable entrepreneurship and will be developed further. For this reason, there is a need for performance indicators at the right strategic level.

The purpose of sustainable entrepreneurship is to incorporate a more strategic environmental management into the ambition of companies to safeguard their continuity and to support individual companies by providing them with jointly developed, appropriate concepts and methods.

This next step includes a swift from the 'negotiated' to the 'induced' phase (and even further on to again 'unconstrained' phase – Chapter 2). A swift from where the government sets targets and objectives, but also makes provisions for negotiations between authorities and industry to the 'induced' approach in which companies which invest and make improvement beyond standard possible are encouraged and rewarded to improve further on. It is important to realise that this phase will bring a decline in direct government influence on the environmental performance of companies, while the influence of other stakeholders will increase.

Our ultimate goal for the next years is not to rely on sustainable entrepreneurs, but to develop and implement business and government policies that have fully integrated the goals of sustainable industrial development. When improvement of environmental (and social) performance is integrated in the market incentives for companies we have the best safeguard for sustainable development.

It is our challenge to turn sustainability from 'responsible' into 'competitive' entrepreneurship.

References

Ministry VROM (1995): Company environmental management as a basis for a different relationship between companies and governmental authorities – A guide for governmental authorities and companies, distribution no. VROM 96222/h/5–96, 20603/201.

Ministry VROM (1997): Environmental management systems. Evaluation 1996, no. 1997–12, distribution no. VROM 97706/h/12–97, 13118/169.

Ministry VROM (1997): Environmental management. A general view. Environmental policy of The Netherlands, distribution no. VROM 97170/h/4–97, 22512/210.

Ministry VROM et al. (1997): Policy document on Environment and Economy. Towards a sustainable economy, distribution no. VROM 97687/a/1–98.

Ministry VROM (1998): Silent revolution. The story of a remarkable environmental success. Dutch industry and the Dutch government conclude contracts to work together for a better environment, distribution no. VROM 18204/191 (book) and 18205/191 (video-pal), 18206/191 (video-ntsc), and 18207/191 (video-secam).

Ministry VROM et al. (1998): National Environmental Policy Plan 3. The summary, distribution no. VROM 97672/b/3–98, 13093/168.

Samsom, H.D. Tjeenk Willink (1997): The new environmental management systems test, Alphen aan den Rijn, ISBN 90 422 0121 5.

6 Mandatory Environmental Reporting: Chance or Risk for Companies?

The U.S. Toxic Release Inventory and Related Mandatory Reporting and the New Jersey Release and Pollution Prevention Report

Michael Aucott

The U.S. Toxic Release Inventory

The Toxic Release Inventory (TRI) is a database that contains information about the releases of toxic chemicals from U.S. manufacturing facilities into the environment. It also contains information on the amounts of these chemicals recycled, burned for energy recovery, or treated at the facility and shipped to other locations for energy recovery, recycling, treatment, or disposal. This information is required to be submitted to the U.S. Environmental Protection Agency (EPA) each year, and is available to the public.

A facility must report to the TRI if it meets or exceeds certain qualifications and thresholds. The first qualification is that it must be a manufacturing facility that is classified within Standard Industrial Classification (SIC) codes 20 through 39. The SIC system classifies facilities according to their principal business activity, which is usually defined by the product of that activity. SIC codes 20 through 39 represent manufacturing activities as indicated below:

Standard Industrial Classification (SIC) Codes: Industrial Facilities

20 Food and Kindred Products
21 Tobacco Products
22 Textile Mill Products
23 Apparel and Other Finished Products made from Fabrics, etc.
24 Lumber and Wood Products, Except Furniture
25 Furniture and Fixtures
26 Paper and Allied Products
27 Printing, Publishing, and Allied Industries
28 Chemicals and Allied Products
29 Petroleum Refining and Related Industries
30 Rubber and Miscellaneous Plastics Products
31 Leather and Leather Products
32 Stone, Clay, Glass, and Concrete Products
33 Primary Metals Industries
34 Fabricated Metal Products, except Machinery and Transportation Equipment
35 Industrial and Commercial Machinery and Computer Equipment

36 Electronic and Other Electrical Equipment, etc., except Computer Equipment
37 Transportation Equipment
38 Measuring, Analyzing, and Controlling Instruments, etc.
39 Miscellaneous Manufacturing Industries

In addition to the SIC code qualification, facility must also have the equivalent of 10 or more full-time employees. Further, it must manufacture or process more than 25 000 pounds (11 340 kg) of a listed chemical, or use in some other manner more than 10 000 pounds (4536 kg) of a listed chemical. If, however, the facility's total generation of a chemical as production-related waste during the year is less than 500 pounds (227 kg), it does not have to report for that chemical unless it manufactures, processes, or otherwise uses more than 1 000 000 pounds (453 600 kg) of that chemical within the year.

Presently, there are about 650 chemicals on the reporting list. Included are most large production chemicals which have shown evidence of toxicity of one form or another. Carcinogenicity has been a major factor in the inclusion of a chemical to the TRI list, but associations of a chemical with other types of adverse health or environmental impact has also led to its inclusion to the list. For example, all of chlorofluorocarbon and related ozone-depleting chemicals banned under the Montreal Protocol and subsequent international agreements, and most of the hydrofluorochlorocarbon alternatives to these chemicals, were recently added to the TRI list.

For each chemical above the threshold quantities (described above), the facility must report the following:

- Amounts released to the air (either as a fugitive or stack emission), to surface water, to underground injection, or to land.
- Amounts transferred off-site to recycling facilities, to sewage treatment plants, to other types of treatment facilities, to disposal facilities, or for energy recovery.
- Amounts managed on-site by recycling, by energy recovery, or by treatment.

The facility must also indicate the approximate maximum amount of the chemical on-site during the year, these types of activities conducted at the facility involving the chemical, and activities undertaken to prevent pollution and waste generation. Other information about the facility must also be provided, including the environmental permits held, and identifying information such as the name and telephone number of a contact person.

The TRI began with the passage, in 1986, of the Emergency Planning and Community Right-to-Know Act. Much of the impetus for the law was the occurrence of several large accidents involving chemicals, including the one at Bhopal, India in 1984 which killed and injured thousands of people and several similar, though much smaller incidents in the U.S. Momentum had already been building at the local level in the U.S. for disclosure of information on chemicals and pollution. By 1986, approximately 30 states or cities had some form of pollution disclosure requirements on the books (Hearne 1996).

During the debate prior to the passage of the TRI law, many industry representatives argued strongly against the concept of public reporting. They expressed concerns about the possibility of public over-reaction, arguing that, although large volumes of chemicals were released in many cases, these releases were legal and thus did not constitute unacceptable risk. Industry officials also voiced worries about the potential loss of trade secrets and confidential business information. Industry sought to restrict access to the TRI information to government health and safety officials (Hearne 1996).

Nevertheless, the TRI law was passed with full disclosure provisions intact. The first reports covered calendar year 1987, and were made public in the spring of 1989. Indeed, the large quantities of chemicals released as wastes directly to the environment or shipped to off-site facilities were surprising to many, and made for headline news throughout the U.S. So far, however, there has been no indication that the TRI has released confidential information or led to extreme public reactions.

There have been numerous changes to the TRI since 1987. Some changes have been deletions of chemicals. Citizens, including industry, may petition the U.S. EPA to add or remove chemicals, and there has been a very active petitions process since the beginning (Sasnett 1994). Recently, several high-volume chemicals were removed from the list, and others were given different thresholds or qualified in such a way that much reporting of them is no longer required. Reporting of all chemicals generated as wastes below 500 pounds (unless more than 1 000 000 pounds are used at the facility) was recently eliminated entirely.

Additions have also occurred. In 1990, the TRI's cope was expanded by the passage of the U.S. Pollution Prevention Act, which required the addition of reporting to cover the quantities shipped off-site for recycling or energy recovery or managed on-site with these methods. In 1994, nearly 300 chemicals, most of them pesticides, were added to the list, effective for reporting year 1995. The EPA has also announced plans to expand the list of covered facilities beyond the manufacturing sector to perhaps include mines, utilities, airports, waste treatment facilities, oil and gas exploration and production facilities, and freight and warehousing facilities. (Sasnett 1994). None of these additions or deletions have been without controversy, some of it intense. Recently, the addition of the approximately 300 chemicals became the subject of litigation.

One of the most interesting and potentially important of EPA's proposed additions to the TRI is inclusion of throughput, i.e., mass balance, or input/output, reporting. The potential benefits, and costs, of such an addition have prompted considerable discussion, much of it quite polarized. Germane to a discussion of mandatory reporting of throughput information are experiences of jurisdictions where such reporting is in place. In the U.S., two states require throughput reporting. One of these is Massachusetts; the other is New Jersey.

The New Jersey Release and Pollution Prevention Report and Pollution Prevention Rules

New Jersey is one of the smallest of the United States, with a land area of 7400 square miles (19 200 sq km). It is densely populated, with 8 million people, and heavily industrialized. The chemical industry is especially well-represented in New Jersey. In the U.S. in 1991, New Jersey was exceeded only by Texas in total value of chemical shipments, with a total of over 24 billion dollars worth. It also led the U.S. in chemical employment, with over 100 thousand workers (CMA 1995). Establishments producing pharmaceuticals, botanicals, and specialty chemicals such as flavors and fragrances make up a significant portion of New Jersey's chemical industry (Aucott et al. 1994).

New Jersey also has more hazardous waste sites on the U.S. national priority list (based on an assessment of the need for remediation) than any other state. The state has also had its share of fires, explosions, and releases from manufacturing and waste management facilities in the past. Driven in part by some of the state's adverse experiences with toxic chemicals, the New Jersey legislature passed the Worker and Community Right-to-Know Act. The rules pursuant to this law require not only the reporting of releases and transfers of toxic chemicals, but their inputs and outputs, or throughput, as well. These rules became effective in 1984, with the first reports covering calendar year 1987.

New Jersey's throughput reporting law covers the same group of industrial facilities and the same list of chemicals as the TRI. But, in addition to the data elements of the TRI, New Jersey requires facilities to report the following inputs and outputs for each chemical used above 10 000 pounds per year:

- Starting inventory
- Amount brought on site
- Amount produced on site
- Amount consumed (i.e., chemically changed into another compound) on-site
- Amount shipped off-site as or in product
- Ending inventory

These reporting items, along with amounts released and amounts managed on and off-site through recycling, energy recovery, treatment, and disposal, permit a crude mass balance of reported chemicals at the facility level. Two quantities, use and nonproduct output, can be derived from these data. Use constitutes the starting inventory plus the amounts produced on site and brought on site, minus the ending inventory. Nonproduct output represents any amount which leaves a process on-site which is not product, and thus represents all production-related wastes prior to any form of treatment, recycling, or other processing.

In 1991, New Jersey passed the Pollution Prevention Act. This law builds on the concept of the mass balance, but extends it to a process level. Under this law, covered facilities, which are the same group which must report under the TRI, are required to develop pollution prevention plans. During the development of a pollution prevention plan, facilities must identify the inputs and outputs of each process which uses listed chemicals, and identify options for reducing either the use of the toxic chemical, or its production as nonproduct output. The law does not require that any specific reduction goal be developed or achieved, but it does require companies to submit a publicly-available summary of their pollution prevention plan. The summary identifies processes which use listed chemicals, and reports on the reduction goals and methods which the facility intends to use to reach these goals. Companies must also annually report their progress relative to the goals.

Benefits and Costs (Chances and Risks) of Mandatory Reporting Programs

There are many factors to consider when attempting to evaluate the overall effect of mandatory reporting programs such as the U.S. TRI and New Jersey's throughput reporting. One must look at not only the benefits and costs to a particular company or facility, but at the overall economic, environmental, and human health impacts. A significant limitation of any evaluation is that these programs have a very short history. These concerns are even more pronounced regarding New Jersey's pollution prevention program, which has only been in effect since 1993.

One of the major hopes of the mandatory reporting concept is that reporting requirements will stimulate companies to take a closer look at those items which must be reported, and that this closer look will in turn stimulate better management. There is also a goal, expressed in the law which created the TRI, that the broad availability of the data will "assist government agencies, researchers, and other persons in the conduct of research and data gathering; to aid in the development of appropriate regulations, guidelines, and standards; and for other similar purposes." (EPCRA 1986)

Clearly, from the standpoint of an individual facility or company, TRI data offer the chance for good publicity. Relatively low reported quantities, or significant progress in reducing quantities reported, can bolster a company's image as a good corporate citizen. There is little doubt that, for many companies, a good image can ultimately lead to higher profits. To the degree that mandatory reporting fosters the collection and utilization of better data on sources of wastes, inefficiencies in production processes, and relative contributions of different production processes to a companies costs of environmental compliance such reporting can help a company save money and avoid potential future costs or liabilities (such as for waste cleanup costs at an off-site facility receiving a company's wastes).

Conversely, negative publicity can result if a company appears high on a list of releasers of wastes. There are also the not insignificant costs of preparing the necessary reports and collecting the necessary information. Further, if confidential or proprietary information is revealed that is useful to a company's competitors, a company could lose market share.

From a broader, societal perspective, the information obtained through mandatory industrial reporting, if it is good information and readily available, offers the hope of leading eventually to more appropriate and cost-effective regulations and better overall management of industrial chemicals. On the other hand, if the mandatory reporting focuses on inappropriate items, or the data is poorly presented, false impressions of progress, inflated perceptions of risk, and general public confusion could result.

It is still too early for a definitive picture of the net cost/benefit balance of the TRI, New Jersey's throughput reporting, New Jersey's pollution prevention rules, and similar reporting programs (such as that of Massachusetts), and of the mandatory reporting concept in general. However, there are indications that, so far, positive results outweigh the negative.

U.S. and New Jersey Experiences with Release and Throughput Reporting

Decreases in Releases and Transfers

At a national level, it is clear that, since the inception of the TRI, there has been a dramatic reduction in quantities of toxic chemicals released to the environment and transferred off-site to waste treatment and disposal facilities. Moreover, this reduction shows a steady trend. Releases and transfers of all listed chemicals dropped from approximately 6.5 billion pounds in 1988 to less than 4 billion pounds in 1993 (US EPA 1995). Similar drops in TRI quantities show up for most subsets of chemicals analyzed and for most states. New Jersey data is illustrative of this downward trend (Fig. 1).

In Fig. 1, the dotted lines represent exponential trends determined by computer. As is clear from the picture, they fit the actual data (dark squares) quite well, and reveal steady decreases in both transfers and releases.

Some Support from Industry

Also, some in industry, at first largely opposed to the TRI, have become supporters, at least of the TRI as it was before the most recent addition of chemicals. The chemical industry, for instance, has stated its support of the TRI because of its commitment to pollution prevention, and because it is clear that the TRI has provided a measure of industry's continuing success in reducing releases of waste materials from its facilities (Ekart 1994). The credibility of the TRI was cited by one official from a large U.S. chemical firm (Monsanto), who stated "public disclosure of the

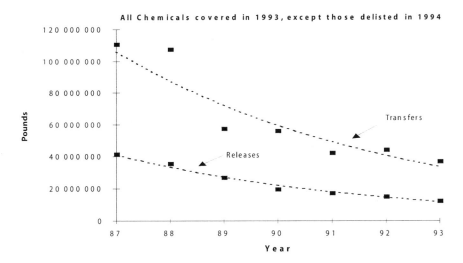

Fig. 1.   New Jersey TRI releases and transfers

TRI has been a powerful motivator to companies. to increase our efforts to reduce emissions. The TRI provides a means where the public can track our progress and do so on a consistent, measurable basis. We are convinced that this activity will ultimately result in cost savings for the company and a competitive advantage." (Hearne 1996,) (However, the chemical industry opposes the recent addition of new chemicals to the TRI and proposed further expansions.)

Costs of Reporting

Industry continues to be concerned about the costs of reporting. The U.S. EPA has estimated that the cost to covered facilities of TRI reporting was about 238 million dollars in 1993, and that this cost will rise to over 400 million before leveling off at about 320 million dollars in the mid 1990's (CMA 1995). This figure is based on EPA's estimate that it takes about 43 hours of a company's time to prepare a TRI report for one chemical. Additional time needed to prepare New Jersey' through-put data has been estimated at approximately 22.5 hours per chemical. (Opperman 1995). One study, however, found that the actual time (in addition to that spent to prepare the TRI report) needed by firms to prepare New Jersey's throughput report was closer to 7.5 hours per chemical (Hearne 1993). The additional costs of completing the throughput report may be less for companies that use the mass balance approach in completing their TRI; there is evidence that, in fact, some companies complete the New Jersey throughput first, and then use these data to help complete the TRI (Opperman 1995). Regardless, the total costs of TRI and throughput reporting appear small compared to the estimated total costs of

U.S. federal environmental regulations of approximately 160 billion dollars per year (Hopkins 1992). Also, there is evidence that companies find mass balance information useful in developing TRI reporting.

Confidentiality Issues

Another of industry's concerns, the potential revelation of confidential information, has not been demonstrated to be a problem with the TRI data. In New Jersey, even with its more detailed and potentially sensitive throughput reporting requirement, confidentiality issues have not been a concern for most reporting facilities either. New Jersey law permits a facility to withhold confidential information if it wishes. Confidential claims are not normally challenged by New Jersey. Nevertheless, for every year since 1987, less than 1.5% of all chemical records have been claimed as confidential.

Trends in Nonproduct Output Generation

As discussed above, there has been a marked downward trend for both the U.S. and New Jersey (and other states) in releases and transfers of toxic chemicals. In the related generation of nonproduct output, however, the U.S. and New Jersey show different trend patterns. Nonproduct outputs are all those outputs of a production process which are not products. They are thus, essentially, all production-related wastes, measured before any reductions or changes due to treatment or disposal. The mass balance concept permits the determination of nonproduct output, and because it encourages a view of all inputs and outputs, can serve to highlight nonproduct outputs from processes which might be ignored if only ultimate releases and transfers were considered. At the national level, the reporting of nonproduct output did not begin until 1991, with the advent of TRI reporting of quantities recycled, treated, and used for energy recovery on-site and sent off site for recycling and energy recovery as well as treatment and disposal. These quantities, closely related to the overall efficiency of a production process, appear in Section 8 of the TRI, and are termed "production-related wastes" by the EPA.

At the national level, there has not been a downward trend in nonproduct output, as the Table 1 illustrates.

In New Jersey, however, the generation of nonproduct output (production-related wastes) shows a consistent decline. Figure 2 shows the nonproduct output of the 5 SIC groups which account for most of the state's nonproduct output for a subset of the TRI chemicals that is termed in New Jersey the Environmental Hazardous Chemicals (EHS chemicals) List.

This decline of New Jersey's nonproduct is just as apparent when the larger group of all TRI chemicals and facilities is viewed. Nonproduct output, of course, is not the only output from a processes. Other outputs include quantities shipped as or in products and amounts actually consumed in the process. Inputs are the amounts

Table 1. Total production-related wastes, all U.S. TRI facilities; billions of pouns; includes all chemicals covered in 1993; 1990 quatities from 1991 data

	1990	1991	1992	1993
Releases	3.83	3.61	3.37	3.16
On-site Energy Recovery	2.88	3.12	3.01	2.82
Off-site Energy Recovery	0.41	0.47	0.47	0.5
On-site Recycling	11.76	12.5	12.18	13.01
Off-site Recycling	2.27	2.98	3.47	3.32
On-site Treatment	9.05	9.43	10.02	9.74
Off-site Treatment	0.66	0.7	0.67	0.64
Totals	30.86	32.81	33.19	33.19

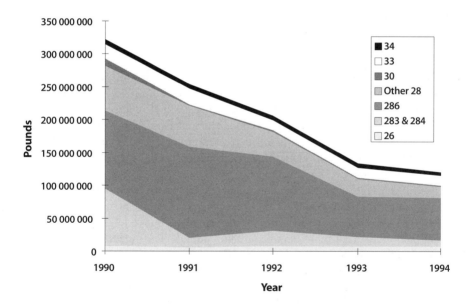

Fig. 2. Nonproduct output: by SIC; New Jersey EHS (RTK) chemicals

brought on site and produced on site. Total use of a chemical (as discussed above) can be defined as the amount brought on site plus the amount produced on site, corrected for changes in inventory from year to year. In other words,

total use = starting inventory + brought on site + produced on site
— ending inventory.

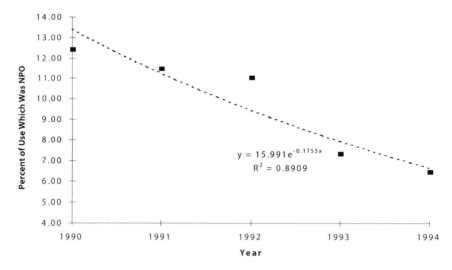

Fig. 3.   NPO as percent of use; RTK (EHS) chemicals only; not including chemicals delisted, etc.
in 1994 (see text); 5 SIC codes only

Nonproduct output represents a percentage of total use, and a decreasing percentage of nonproduct output could be interpreted as a broad indication of increasing efficiency in the use of a chemical (Fig. 3). In New Jersey, a look at nonproduct output of the same subset of chemicals considered above, (which does not
all of the TRI chemicals such as those delisted in 1994) for the same 5 SIC codes
(i.e., 26, 28, 30, 33, and 34) shows that nonproduct output as a percent of use has
shown a consistent decline since 1990.

New Jersey's rate of decline of nonproduct output appears to be among the highest
of U.S. states. The nonproduct output (from the production-related wastes section 8
of the TRI) was determined for all U.S. states that had a 1990 total (as determined
from the 1991 data) of more than 500 million pounds. The quantities were plotted
for each of the years 1990 through 1993, and then the mean percentage change per year
was determined using an exponential regression done by computer. Some states
showed an increasing trend, and some showed a declining trend, as Fig. 4 indicates.

As discussed earlier, the quantities of total U.S. TRI releases and transfers have
shown a steady decline since at least 1988. It has been suggested that this decline is
a result of the operation of a principle that what must be reported will be measured, and that what is measured will be managed.

If this principle is valid, it could be expected that a requirement to report
nonproduct output would lead to better management of this quantity, and, to the
extent that reductions were possible, to reductions. Since New Jersey has been requiring throughput data since 1987, and since nonproduct output quantities are
readily derived from accurate throughput data, it may be that New Jersey's greater
rate of decline of nonproduct output is, in fact, related to the state's requirement of

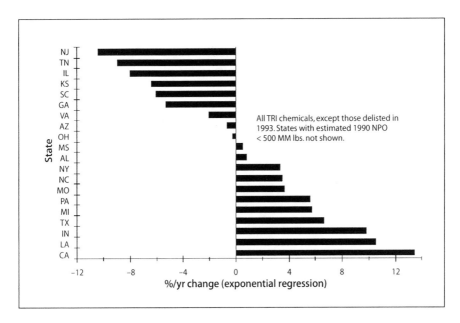

Fig. 4. Trends in NPO generation; from US TRI data 1991–1993

throughput reporting. Supporting this conclusion is the fact that Massachusetts, the only other state that has consistently required throughput reporting, also has a high rate of nonproduct output decline (these data are not shown, however, because the state's total nonproduct output is below 500 million pounds).

On the other hand, the decline in New Jersey could be due to other factors. One possibility is the state's history of strict enforcement of relatively vigorous environmental laws. Arguing against this reason, however, is California's low position in Fig. 4. It is likely that California ranks with New Jersey near the top among U.S. states in vigorousness of environmental laws, yet it is at the bottom in nonproduct output trend, with an average increase of more than 12 percent per year.

It could also be that the New Jersey nonproduct output decline is the result of declining business activity. In fact, New Jersey has endured an economic downturn in recent years of greater severity than most states. However, when economic performance, as determined from U.S. Census Bureau reports of value added for all industrial sectors per year per state (U.S. Dept. Commerce 1991, 1992), is compared with the nonproduct output trends as discussed above, there appears to be little, if any correlation, as Fig. 5 indicates.

In Fig. 5 each state is represented by a dark square. Value added was corrected for inflation, and trends were derived from an exponential regression in the same manner as the nonproduct output trends were derived. The linear regression line shown does not fit the data well, although there may be some slight relation. New Jersey, with an average decline in value added over the period indicated of about

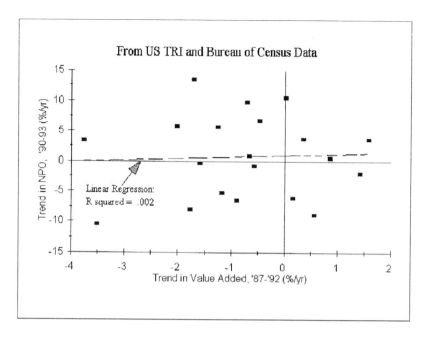

Fig. 5.   Value added versus NPO Trends; by state

3.5% and an average nonproduct output decline of over 10%, is represented by the square in the lower left.

A closer look was taken at the possible relation between trends in nonproduct output generation and value added. In New Jersey, most of the decline in nonproduct output has been accomplished by facilities in SIC code 28, the chemical manufacturers (see Fig. 2). In fact, this SIC group has shown an upward trend in value added in New Jersey over the last 10 or more years, as Fig. 6 indicates.

These data indicate that, perhaps, an inverse relationship may exist between value added and nonproduct output trends. In fact, when the economic performance of other SIC codes in both New Jersey and the U.S. as a whole is compared with these groups' nonproduct output trends, an inverse relation also is apparent. Research in this area is preliminary, and more aspects of this possible inverse relation must be analyzed. However, it may be that the often repeated maxim that good business and pollution prevention go together is supported by these data.

Total Use Quantities Often Dwarf Nonproduct Output Quantities

In addition to trends in nonproduct output discussed above, other types of information relevant to the overall management of chemicals emerges from a review of New Jersey's throughput data. One such type of information is the often surprising

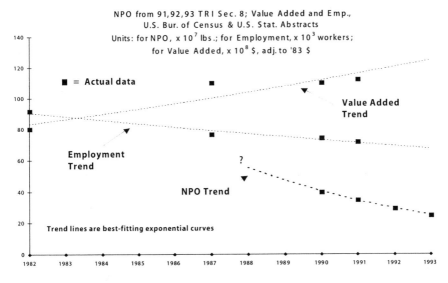

Fig. 6. New Jersey SIC 28 trends: value added, employment and NPO

quantities of chemicals used by industrial facilities relative to the quantities of nonproduct output produced (Aucott 1994) (Releases and transfers, which are subsets of nonproduct output, are smaller still in relation to total use.) Table 2 based on 1994 New Jersey throughput data, shows approximate total use quantities, and the approximate percent of this total use which is either shipped off site as or in product, consumed on site, or which leaves processes as nonproduct output.

With many chemicals, much of the quantity shipped off-site as or in product will eventually reach the environment, sometimes in situations where significant human exposure can result. This is the case with dichloromethane (methylene chloride). Much of this material that is shipped as or in product is used in paint strippers, and vaporizes as this product is used.

Table 2. Emissions and transfers in relation to total use

Chemical	Used (Lbs.)	% Consumed	% Nonproduct Output	% Shipped as or in Product
Cadmium	840 000	0	8	92
Chlorine	103 000 000	~99	<1	~1
DEHP	38 000 000	4	~1	95
Dichloromethane	8 800 000	<1	61	38
Perchloroethylene	340 000	20	40	40

Considerations Regarding Possible Global Release/Transfer and Throughput Reporting

Although the short time they have been in use makes it impossible to definitively evaluate the cost and benefits of mandatory reporting of releases and transfers and throughput, there appears to be significant value to these programs. Releases and transfers have shown a marked decline in the U.S., almost certainly due, at least in part, the TRI. Companies have changed from opposition to the TRI to support in some cases, because they have seen that reductions in releases, etc. have positive benefits. In New Jersey, nonproduct output, a broader measure of production-related wastes (and thus, in a sense, of inefficiencies) has shown a decline. Confidentiality concerns, judged by the low frequency of confidential claims in New Jersey, appear to be manageable. The paperwork burdens of these mandatory reporting programs, although not insignificant, appear small relative to other regulatory burdens faced by companies. The detailed picture afforded to companies by a mass balance approach to their processes appears useful in stimulating increased efficiency (Epstein 1996) The broad picture of chemical use in society afforded by throughput reporting offers the hope of eventual improvements in environmental management and regulation.

Based in part on these ideas, several general recommendations can be made regarding the possible extension of release and transfer and throughput reporting to a global level.

- Focus on truly important chemicals
- Include all significant sources
- Adopt optimum-sized thresholds
- Include throughput reporting to avoid major omissions and an unbalanced picture of chemical uses and ultimate emissions
- Provide confidentiality protection
- Couple the inception of release/transfer and throughput reporting to the elimination of other, less useful reporting programs
- Combine these mandatory reports with measures which permit the true assessment of environmental and human health impacts, including chemicals' transport, fate, and toxicity.

References

Aucott, M. (1994/95): Expanding the Reach of the TRI – New Challenges for Industry – Releases versus Throughput, Pollution Prevention Review, Vol. 5, No. 1.

Aucott, M.; Healey, M.; Ballantine, T. (1994): Profile of New Jersey Industry, 1990: Issues Relating to Pollution Prevention, New Jersey Department of Environmental Protection, Trenton, NJ.

Chemical Manufacturers Association (CMA) (1995): U.S. Chemical Industry Statistical Handbook; 1995, Chemical Manufacturers Association, 1300 Wilson Blvd., Arlington, VA, 22209.

Ekhart, N. (1994/95): Expanding the Reach of the TRI – New Challenges for Industry – A Chemical Industry Perspective, Pollution Prevention Review, Vol. 5, No. 1, Winter.

Emergency Planning and Community Right-to-Know Act (EPCRA) (1986): *Section 313, subsection (h)*, U.S. Congress, Washington, DC.

Epstein, Marc (1996): Measuring Corporate Environmental Performance, ISBN 0-7863-0230-5, IMA Foundation for Applied Research, Inc., Montvale, NJ.

Hearne, S. (1996,): (Article on TRI and Throughput), Environment Magazine, July/August, in press.

Hearne, S. (1993):Materials Accounting as a Potential Supplement to the Toxics Release Inventory for Pollution Prevention Measurement Purposes, Ph.D. Thesis, Columbia University, NY, NY, and NJ Department of Environmental Protection, Trenton, NJ.

Hopkins, Thomas D. (1992): The Costs of Federal Regulation, Rochester Institute of Technology, Rochester, NY.

Opperman, A. (1995): personal communication, Bureau of Chemical Release Information and Prevention, New Jersey Department of Environmental Protection, Trenton, NJ.

Sasnett, S. (1994/95): Expanding the Reach of the TRI – New Challenges for Industry – An EPA Perspective, Pollution Prevention Review, Vol. 5, No. 1.

United States Department of Commerce (1993): Bureau of the Census, Annual Survey of Manufacturers and Census of Manufacturers, 1991 and 1992. Superintendent of Documents, U.S. Government Printing Office, Washington, DC, 20402.

United States Environmental Protection Agency (US EPA) (1995): 1993 Toxics Release Inventory; Public Data Release, U.S. EPA, Office of Pollution Prevention and Toxics, Washington, DC, 20460.

7 Environmental Performance Evaluation – The Link Between Management Systems and Reality

Tron Kleivane

The Environmental Challenge

Since the term "Sustainable Development" was keyed by the Brundtland Commission in 1987, it has modified the environmental agenda and become a focal point in the political debate. The term has survived and gained momentum even if, or rather because, it is ambiguous and difficult to define. The term is central to policymaking even if it can sound a bit like an old hat when confronted with the serious problems of unemployment, the lack of economic growth and the sosial disrubtance that caracterizes most European countries at the end of the century. The environmental challenge is no fad. The climate issue alone will ensure a global political attention on environment for decades to come and the need for action will grow stronger for every year. The question will never be if but how to promote sustainable development.

On a macro economic level the problem, however, is to develop new and efficient policies adapted to a global, technology driven and strongly interdependent economy.

Rules and regulations alone will not ensure the desired results both because of the complexity of the economy and because of the limited effects of national legislation on the international scene. In a period of time with rising unemployement and stronger international competition, the legislative approach seems rather insufficient. Most experts agree that the environment has to be internalized as a valuable good in the market economy. What is needed is a much stronger activation of the market mechanisms. It must become profitable to act according to the needs of the environment. It should pay to be proactive. The internalization of environmental externalities in the market is probably the only way to ensure sustainable development.

All markets depend upon reliable, relevant, quantifiable and consistent information, generally expressed in prices. There is no consistent environmental information available in todays marketplace. We do not yet know how to price the environment and we have not yet delveloped alternative quantifiable information.

On a company level there is an increasing and continued environmental concern arising from external and internal interested parties wanting to influence the way business and industry is dealing with environmental issues. The variety and quality of stakeholder pressure has enforced top management in companies to work harder to get on top of the environmental agenda. What are the real environmental challenges for the company, now and in the future? What are the related financial risks? How to define the companys policies, objectives and targets?

From being reactive, protective and mostly concerned by the hassle and costs occuring from stakeholder pressure, the companies are now increasingly seeking ways to understand, improve and demonstrate sound environmental performance by controlling the impact of their activities, products and services on the environment.

The development of standardized management systems, tools and guidelines for self improvement is seen as fundamental in order to foster proactivity in business and avoid unnecessary or counterproductive stop-go regulations. An ever increasing environmental awareness both in industry and in society at large will put more demands for meaningful information on management. Either management develops tools and information to be able to participate on equal terms in the environmental debate, or the various stakeholders will set their agendas imposing reactive measures from a management in loss of control.

Consistent, relevant and comparable environmental performance information has become crucial for management control and is seen as a prerequisite for long term company health and financial results.

Information on environmental performance is thus becoming a strategic issue both in a macro economic perspective and on a company level. But what exactly is performance information and how would such information be created? Today there is an undeclared battle going on in the marketplace between various stakeholders seeking to influence the environmetal agenda through their selection of presented information. Participants in this battle are companies, business organisations, environmental NGOs, scientific experts, financial stakeholders, the media and public autorities, just to name some of the most important players. The result is still a rather complicated public debate with questionable quality. Even if there are signs of improvements, the lack of a common language is still impeding on the quality of the debate and on real progress towards sustainable development.

The question is how to transform masses of data into meaningful information.

The answer from a business perspective is Environmental Performance Indicators (EPIs). Indicators are only representing parts of reality. Still, the right indicator will give better insight in reality than the compilation of all available data. By selecting appropriate indicators, a company will be in a position to measure, assess and communicate the organisations environmental performance. For management, indicators are means to move from masses of partly inconsistent and irrelevant environmental data, to meaningfull and quantitative environmental information.

The History of Environmental Performance Evaluation

Several initiatives have been launched to help companies develop and use environmental information in integration with current business needs. Some of the most important initiatives these last ten years have been the ICC Business Charter for Sustainable Development, GEMI's Environmental Self Assessement Program; European Green Table's EPI-project and ISOs work on environmental management standards.

The work of ISO on Environmental Management (ISO/TC 207), initiated in 1990 by the Business Council for Sustainable Development (now WBCSD) and the European Green Table, is a focal forum for the development of tools and systems aiming at establishing standards for sound environmental management in business and industry.

The term "environmental performance evaluation" (EPE) was developed by ISO SAGE in the preparation of the standard writing process on environmental management in TC 207. It was put forward by proactive countries and business communities to emphasise the importance of environmental impact assessement. Continuous improvement on a system level alone (EMS) would not be sufficient for management to be on top of the environmental agenda and meet stakeholder requirements. There would be a need to get deeper into the realities of the interaction between the companies and the environment.

The establishment of standards for environmental performance evaluation has been the scope of work in subcommittee 4 (SC4) in TC 207 and is probably one of the most challenging tasks within the TC 207. How can the organisation define, describe, analyze, measure and evaluate the significant environmental aspects of its activities? The establishment of such standards will have far reaching consequences. In many ways it can be compared with the development of financial accounting systems in the previous century. It took more than fifty years to translate the economic realities in companies into a standardized, consistent and comprehensive accounting system strengthening management control and providing a basis for transparency and stakeholder participation. The challenge today is to contribute to substantially shorten the time needed to translate environmental impacts and concerns into meaningful and consistent performance information to be used by the management both for internal and external purposes.

The standard on EPE, which will be a guideline standard classified ISO 14031, will probably become a Draft International Standard by the end of 1997.

The main purpose of the standard will be to give guidance in selecting and using environmental performance indicators to measure, assess and communicate an organizations environmental performance.

One of the main problems throughout the standard writing process has been the lack of relevant empirical experience to support the development of a practical guideline. Some companies like British Telecom, BP, Shell, Statoil, Rhone Polenc, 3M, Norsk Hydro, Noranda and Novo Nordisk have been using environmental indicators for some years to get a better picture of how various environmental key figures are developing. Most of these companies have chosen indicators to reflect their existing data, to highlight some chosen environmental issues and support internal management initiatives. Few of these companies have chosen a comprehensive approach, suggesting a methodology to sort out which environmental aspects should be considered significant and which aspects should not, and how should these significant environmental aspects be measured and monitored by relevant indicators to reflect the companys environmental performance.

The Methodology – Experiences from the EPI-project

To strengthen the company initiatives and give room for a more systematic approach, the European Green Table launched its EPI-project in 1990, while supporting the establishment of a technical committee in ISO to develop standards for environmental performance evaluation.

The project was finished in march 1997 after having developed and tested a methodology for selecting and using EPIs in 12 pilot companies.

The purpose of the methodology developed in the EPI-project was to help the companies select and define their "significant environmental aspects" through a transparent and verifiable top-down and bottom-up approach based on the companys own best knowledge and judgements. Within the scope of the selected significant environmental aspects the methodology aimed at helping the company define relevant, representative and measurable indicators enabling the company to monitor, report and communicate their environmental performance.

The top-down approach was implemented through classification of environmental issues seen from a broad management perspective at site level. It was aimed at producing meaningfull information for top management for strategic, economic and financial purposes.

The developed methodology was based on a life cycle approach and the interaction between the organizations activities, the environment and stakeholders requirements. The significant environmental aspects of the companies and representative EPIs were defined after conducting an environmental assessement and a stakeholder assessement.

A step by step process linked to the traditionnal plan-do-check-act approach guided the companies in their implementation.

The process leading up to EPIs is briefly outlined and exemplified in Fig. 1.

The testing within the pilot companies confirmed that the developed EPI methodology could help companies to:

- Provide the management with concise and quantifiable environmental information;
- Define their significant environmental aspects and describe and measure their environmental performance;
- Improve the basis for their environmental policy, objectives and targets;
- Improve the basis for their internal and external reporting and communication on environmental issues;
- Focus on, and demonstrate, continual improvement of environmental performance;
- Improve the basis for internal and external benchmarking;
- Prepare for certification according to ISO 14001 and EMAS.

The results from the EPI-project have influenced the standard writing of EPE (14031) on a continual basis, and helped focus on clarity, consistency and practicability.

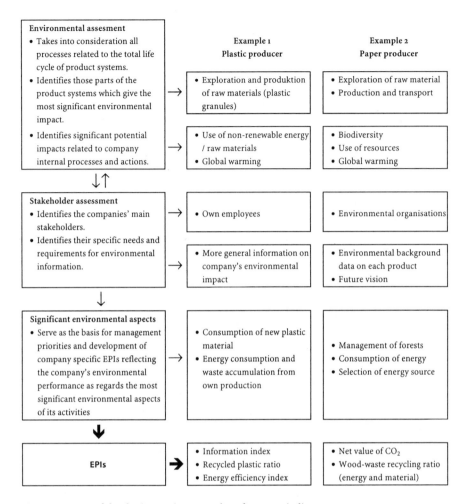

Fig. 1.   Process of developing environmental performance indicators

Unsolved Issues in the Standard Writing of EPE

A new ISO standard giving guidance to organizations wishing to evaluate their environmental performance, will strongly influence the quality, the consistency and the relevance of environmental information both on a company level and for use by stakeholders in a wider perspective.

All standard writing is political and as such subject to a lot of give and take before consensus is reached. Subcommitte 4 responsible for EPE in ISO TC 207 has had a particularly difficult task both because of the scientific challenges in the approach, and because of the lack of empirical experience to start off the standard writing process.

Two areas of discussion have been particularly difficult throughout the process – the choice between systems performance and environmental performance, and the format and purpose of the standard. These areas have influenced on other issues linked to the planning of EPE such as the importance of a stakeholder assessement; the linkage between the managements objectives and targets and EPE i.e. how to define significant environmental aspects and the definition and the content of an initial environmental review.

Some countries have been very preoccupied in limiting the scope of work within SC4 to reflect and submit to the content of the 14001 and 14004 standards for environmental management systems (EMS). According to these standards, however, the definition of "environmental performance" could be limited to describe the quality of the management system as such without describing the results of the organisations interaction with the environment.

Others within SC4 has fought hard to keep a more independent line, arguing that what is really asked for both the companies and from the market is information on performance related to the environment. Either the company produces such information or the stakeholders will do it.

The countries divide in a quite identical manner on the format and purpose of the standard. The countries being "loyal" to the limitations set forward in 14001 are eager to emphasise that 14031 is a guideline standard, and that it should be as open ended as possible to give room for the users own interpretations. These countries want to include several approaches on methodology and a varity of examples and comments to ensure that the user is not mislead to believe there is only one way of evaluating environmental performance. The countries preoccupied with the linkage to the environmental impacts of the companies have a different approach. They want a short and instructive document that can be of practical help even for the small and medium sized companies wanting to evaluate and get control of their environmental performance whether or not they have a formal EMS in place. They want to give priority to clarity and consistency of the conceptual approach arguing that the document is only a guideline to spur organisations development of their own tools.

These differences spill over on other issues as well. The first group of countries take it for granted that in most companies will have an EMS in place and that management will have a predefined set of environmental policies, objectives and targets. Thus EPE should be limited only to give guidance on the technical parts of measuring, assessing and communicating performance using indicators as tools.

For the second group of countries however, most companies would wish to have guidance also to establish their first set of objectives and targets or improve their existing ones. Before a company can establish policies, objectives and targets it has to know where it is and where to go. This has to be done on the basis of information gathered in an initial environmental review as suggested in ISO 14004. The only ISO guidance however, on how to conduct an initial environmental review will be indirectly in the 14031 standard on EPE. Indirectly because the term initial environmental review has not yet been adopted by SC4 to describe the first time a company does an EPE process.

The same dilemma exists with "significant environmental aspects". Are the definiton of these an integral part of the planning of EPE, and an activity where guidance is needed? In the existing draft document 14031 the term is watered down not to disturb countries loyal to the 14001 limitations.

Moreover, experience from the EPI-project suggests that stakeholder assessements are extremely useful in the process of defining significant aspects and selecting indicators. The final standard should emphasise this element.

The quality of the 14031 standard will also to a certain extent define the linkages between ISO and the Eco Management and Audit Scheme (EMAS) put forward by the European Community. EMAS requires regular reporting of continual improvement of environmental performance in a yearly statement. This obligation goes beyond the requirements in ISO 14001. Thus a bridging between ISO 14001 and EMAS is required. The ISO 14031 standard could contribute to bridge the differencies provided proactive attitude in the remaining discussions to be held in SC4.

Some of the differences in view are certainly due to different cultural backgrounds and differences in the countries legal systems. However, there should be no doubt as to what is needed in business and industry. There is a need for clear and concise documents giving guidance to organizations of all types and sizes, helping them to become environmental literates. Government, financial stakeholders, NGOs and customers are increasingly asking for relevant information on the companies environmental performance.

By chosing to support the companies working on evaluating the environmental impacts of their activities, ISO 14031 can strengthen both the environment and industrys agenda power when confronted with the future environmental challenges.

References

BMU/UBA: Bundesumweltministerium, Umweltbundesamt (Hg.) (1997): Leitfaden betriebliche Umweltkennzahlen, Bonn, Berlin.
BMU/UBA: Bundesumweltministerium/Umweltbundesamt (Hg.) (1995): Handbuch Umweltcontrolling, München.
Loew, T.; Hjalmarsdottir, H. (1996): Umweltkennzahlen für das betriebliche Umweltmanagement, Berlin.

8 Eco-efficiency in Banking – From Assessing the Risks to Expanding the Opportunities

Inge Schuhmacher, Gianreto Gamboni

Banks have now recognized that their own performance is increasingly affected by environmental factors. In the lending business, for example, it is important to determine a borrower's exposure to environmental risks in order to raise the quality of the loan portfolio and avoid potential defaults. And on the other side of the coin, a bank can leverage its in-house environmental expertise into new financing opportunities.

In the investment business, one of the aims is to upgrade the quality of information available to financial analysts by including the environmental dimension. This has the additional benefit of making it possible to capitalize on an emerging market trend: many investors, private and institutional alike, have become sensitized to ecological factors. They want their investments to contribute to the solution of global environmental problems while profiting from the value-enhancing effect that eco-efficiency can have on a company's operations. These considerations are leading the major banks, alongside the alternative "green" banks, to view eco-efficiency as a viable business proposition.

For decades the banks remained unaffected by the "greening" of industry. The non-polluting service sector, in contrast to the traditional smokestack industries, generated relatively little controversy. The environmental market was gladly left to a handful of niche players that offered savers an opportunity to extend concessionary terms to ecological projects by forgoing some of their interest. That picture has now changed. By mid 1998 more than 110 banks around the world (including a number of majors) had signed the United Nations Environment Programme's UNEP declaration, "Banking and the Environment, a Statement by Banks on the Environment and Sustainable Development". In doing so they made a public commitment to sustainable development as a fundamental aspect of sound business management and good corporate citizenship. The aim is to balance the interests of this and future generations. The progress of this credo was impressively documented in the Swiss press in spring of 1997. In the space of a few weeks, a number of financial services providers launched new, environmentally-oriented investment products. What was behind this change in attitude?

Why Does a Bank Get Interested in Eco-efficiency?

The concept of "eco-efficiency", like the concept of sustainability, is used not only by environmental organizations but also by companies aware of the importance of optimizing their environmental performance and translating this awareness into practice.

After years of "end-of-pipe" and "end-of-process" solutions, the implementation of environmental standards in production and product design is the goal today. The advantages are obvious: companies that start with the understanding that the claims of the environment are not simply onerous burdens but pose a strategic challenge and thus constitute one of the foundations for long-range success and a sound corporate future will view ecological questions in a fundamentally more positive light.

Eco-efficiency is defined in many ways, but essentially it involves increasing the value added in relation to both the resources consumed and the impact on the environment. The World Business Council for Sustainable Development (WBCSD) has defined seven elements that contribute to eco-efficiency:

1. Reducing the material intensity of goods and services = lowering costs

2. Reducing the energy intensity of goods and services = lowering costs

3. Reducing toxic dispersion = reducing potential and actual liability

4. Enhancing material recyclability = efficient, closed cycles

5. Extending product durability = increasing customer benefits

6. Maximizing use of renewable resources = long-term strategic advantage

7. Increasing the service intensity of goods and services = opening up new areas of activity

The corporate sector's efforts to implement such strategies can have direct economic consequences, and thus have important implications for the financial services industry. Confronting these issues squarely is therefore central to the integration of environmental factors into a bank's business operations.

In the loan business this is directly related to the bank's role as a lender trying to minimize risk. In the investment business it affects the bank's search for secure investments – and especially growth vehicles – for its customers. Eco-efficiency plays an important role in both areas, because it has a direct bearing on a company's liability exposure, profitability and strategic positioning.

The Loan Process is First and Foremost a Risk Assessment Process

"A borrower in the metalworking industry goes bankrupt as a result of adverse market conditions, the high cost of modernizing obsolete production facilities, and management mistakes. Before the factory complex is put up for auction, the bank discovers that the site has been heavily polluted. As a result the value of the real estate has to be reappraised. The contamination turns out to be so serious that the property is essentially worthless. The bank even has to cancel the auction on both economic and legal grounds, and the loan has to be totally written off."

This case shows how potentially devastating conditions that had neither been reflected in the balance sheet nor addressed by management can suddenly emerge to threaten a company's very existence. Costs can arise that can no longer be borne by the polluter and have to be passed on to third parties such as the state, creditor banks and insurance companies.

Environmental risks can create heavy financial burdens for a company that have a negative impact on net worth and earnings value:

- Litigation and fines,
- the cost of giving evidence, demonstrating compliance or defending against unjust charges,
- decline in value of assets (e.g. due to contamination),
- inadequate or non-existent insurance coverage, higher insurance premiums,
- damage to image leading to loss of trust and diminished market acceptance,
- in the extreme case, going out of business.

Neglecting environmental principles can therefore pose a serious threat to a company's earnings power and shareholder value. Environmental risks of this kind must be counted among the borrower's operating risks, and thus become direct or indirect risks for the bank as well:

- Credit risks: Impaired borrower solvency due to unforeseen major investments required under regulatory or judicial clean-up orders; insurance deductibles in damage settlements; loss of market share or liabilities to third parties;
- Risks to collateral where property put up as security for a loan has been contaminated;
- Unexpected escalation in value adjustments, for instance if environmental damage is discovered during a restructuring phase;
- Exposure to liability in connection with auctions of contaminated factory sites, participation or influence in management, or other involvement (e.g. subsidiaries, cross-holdings);
- Potential loss of image due to financing of environmentally and/or ethically questionable production or projects;
- Risks as financial advisers or consultants, e.g. in mergers and acquisitions deals.

The above factors make it clear that ecological criteria are incorporated into the credit process not only on ethical grounds or out of concern for the bank's image but also because this helps to manage the risks on the loan portfolio. In the US and UK, the laws on lender liability have been instrumental in these developments. In Western Europe the general focus is more on how to protect and preserve the value of collateral in real estate financing and how to achieve sustained earnings value in the long run. The advantage is easy to see: lenders able to assess the interrelated environmental and business risks facing their customers will be holding three trump cards:

- Lower costs, because a portfolio of higher-quality customers requires less provisioning;
- The ability to offer competitive terms tailored to the customer's own risk situation;
- A growing store of expertise in environmental questions that can be employed to develop new financing business.

Swiss Bank Corporation and the other Swiss big banks have drawn the appropriate conclusions and have expanded their credit risk management function to address these aspects. At Swiss Bank Corporation this includes issuing additional directives and credit policy guidelines, defining effective assessment and control procedures, developing tools and techniques for the identification and evaluation of environmental risks (checklists and industrial classifications), establishing an Environmental Management Services unit and providing credit and loan officers with the necessary supplementary training.

Experience has already demonstrated that the bank is not the only beneficiary of this policy, which improves the quality of its customer loan portfolio and reduces the need for provisions. For the borrower, eco-efficiency and a professional approach to environmental management mean lower overall exposure to risk. With the banks now shifting to "risk-adjusted" loan pricing, it can safely be assumed that competent environmental management will be appropriately valued or rewarded by lenders wherever ecological risks make up a significant part of the total risk.

However, a hands-on approach to environmental risks also represents an opportunity for instance, to provide advisory and referral services for customers when problems arise. This can help the customer take preventive action more effectively and handle clean-ups more successfully. The bank's credit risks are reduced and ties with customers are strengthened. Moreover the bank becomes a competent partner for discussions on environmental questions in connection with new financings, and is better to move into the growth markets in this field.

The Key to Investment is Weighing the Opportunities: Ecology in the Investment Business

The principle is simple: investors put up their money because they hope to make a profit. This applies to private investors, pension plans and investment funds alike.

There is no longer any doubt that environmental factors can have a strong influence on a company's current and future credit standing.

For one thing, ecological risk can have a negative impact on the balance sheet. This is particularly true if real estate makes up a large part of a company's assets. The value of land and buildings is immediately and sharply affected whenever a contamination problem comes to light. Potential liabilities can also be created by pollution from production or products, and other threats include the possibility of a loss of image, a negative market reaction or blunders by management.

Steps have already been taken to see that financial analysts and investors receive information on these points. The New York Stock Exchange Commission, for example, requires companies to publish their financial obligation associated with activities that pose a hazard to the environment. The European Union has already issued a directive requiring publication of data relevant to the environment. The obligatory reporting of toxic emissions by analogy with the US Toxic Release Inventory is also being discussed. Nevertheless, the ability of the financial markets to evaluate ecological influences is still in its infancy.

Since future appreciation in value and good earnings prospects are key arguments in an investment decision, all factors that can affect shareholder value – including environmental factors – must be taking into consideration. A common criticism of the financial markets (and especially of financial analysts) is that only short-term factors such as quarterly earnings are examined in their assessments of shareholder value. This is understandable when you realise how little information on long-range corporate goals and strategies is available. The reason is simply that there was little demand for such information in the past, especially with regard to environmental questions, and procedures for evaluating a company's ecological-cum-ecological efficiency were lacking.

In principle, shareholder value is exposed to both short-term and long-term influences. In the short term they have a positive effect on dividends, and in the long run they secure sustained growth in shareholder value.

But what are these internal and external determinants of shareholder value that are connected with eco-efficiency?

Shareholder value	Determining factors	
Stock price rises ↑ Long-term influence	• Strategy • R&D • Corporate culture	• Industry standard • Social / political trends • Regulation • NGOs
High quarterly dividends ↑ Short-term influence	• Quality of costs and workforce • Compliance • Image • Product / quality	• Competition • Regulation • Enforcement agencies • Customers / consumers
	Internal factors	External factors

Fig. 1. Environmental factors influencing shareholder value

Internal factors with a direct influence on shareholder value include eco-efficient product design and quality, compliance with legal requirements, strong management, and so on. *External* factors with an immediate impact include regulatory agencies, customers, competitors and the media (public opinion).

Long-term internal factors affecting shareholder value include the investment policy with regard to eco-efficient technologies and research and development activities. It is vital to motivate the staff with a clear management endorsement of environmental aims and to provide programs to sensitize employees to the issues. Long-term *external* influences include industry standards, environmental legislation, social and political trends related to ecology and environmental organizations.

This illustrates the main levels at which eco-efficiency works: a short-term influence arises from management capabilities primarily in the area of process and product optimization; the long-term influence, on the other hand, stems from strategies and their implementation, as well as the identification of changes in the environment and of new possibilities, for instance benchmarking trends in industry.

Is There any Empirical Evidence of Environmental Factors Moving the Share Price?

The financial markets are more easily convinced if you can show a direct correlation between environmental conduct on the part of a company and its shareholder value. Massive plunges in share prices can be observed in the wake of environmental disasters, such as the Sandoz fire or the Exxon Valdez oil spill. The case of a Swiss firm, Von Roll, provides another dramatic example: when it became known in the run-up to a planned merger with another steel company that years of pollution would mean very high remediation costs, the market value of Von Roll dropped by CHF 83.6 million in just one day as the stock price took a dive from CHF 26 to CHF 22.25. Enormous environmental problems became evident as Von Moos was in the process of taking over Von Roll. In September of 1996, it emerged that environmental regulations had been violated for years although remedies would have been available. Protection of air and water quality and noise abatement procedures were all inadequate. The planned investment in environmental technology now comes to CHF 30 million, plus up to CHF 50 million in clean-up costs.

The financial markets were efficiently reflecting the clean-up costs which the company would have to meet.

Such price setbacks are normally submerged in other developments fairly quickly, making it hard to demonstrate a long-term influence. Depending on the time frame selected, either statement can be proven: environmental performance has an impact on the share price, or it has none.

A more academically disciplined approach to identify a correlation has been taken by empirical studies (listed in the appendix). Particularly in the USA, a large

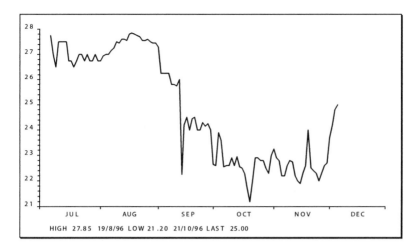

Fig. 2.   Von Roll share price (daily movement from 5 July to 5 December 1996)

number of studies have been carried out over the last three years, some focusing on individual cases and others on comparisons between corporate groups with active or passive environmental management systems. The classification was made partly with the help of quantitative criteria such as emission statistics from the Toxic Release Inventory (TRI), the number and scale of accidents in the chemical industry or incidence of special wastes, and emissions of greenhouse gases. Other factors were more qualitative, such as the degree of commitment to environmental causes as evidenced by the signing of the CERES principles. Most of the studies show that the financial performance of the environmentally active groups is superior, as time goes on, to that of the "laggards". However, it must also be pointed out that long-term comparisons are rare, and also that the positive performance of the companies concerned cannot be ascribed solely to environmental factors.

Instead of ex-post analysis, a qualitative ex-ante approach would appear to make sense, to demonstrate how the "shareholder value drivers" such as strategy, business operations, investment and financing can be affected by corporate environmental behaviour. This was the solution chosen by the WBCSD working group on environmental performance and shareholder value. The participants, from industrial and financial services firms, examined the question of how environmental business conduct affected a company's shareholder value. The starting point consisted of numerous studies confirming that financial analysts paid little heed to environmental factors. The goal of the working group was to establish a dialogue between the corporate sector and the financial markets.

For the dialogue to function, the various players have to adapt their environmental communications to the needs of the target audience.

Existing environmental performance reports are hardly designed to meet the needs of the analysts, because it is difficult to use the reports in their present form

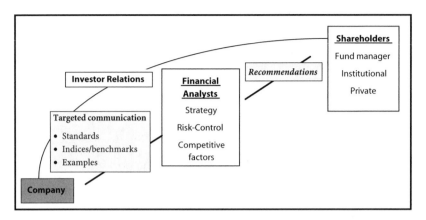

Fig. 3. Model of dialogue for integration of ecological criteria into financial analysis and investment decisions

for the assessment of eco-efficiency (let alone shareholder value). For economic and environmental performance to be gauged, standards have to be worked out and performance indices for specific industries developed. Instead of column after column of environmental data accompanied by 30–70 pages of text, the material should be aggregated and presented as part of the corporate annual report. This will require co-ordination between trade associations and the financial markets. Industry trade groups will have to develop reporting standards and performance indicators compatible with their particular environmental situation. In addition, accountants and financial services professionals should come to an agreement on a consistent format and requirements for the presentation of corporate environmental information. Positive examples of this are the ACBE proposal (ACBE 1997)and the VfU standard for environmental reporting (VfU 1996) by financial services firms.

The Swiss banks, with a working group of ecological experts and financial analysts from the Swiss Bankers Association "Banks and the Environment" working group, have therefore produced a discussion paper on the banks' requirements with regard to environmental reporting. This represents a standard to facilitate the integration of environmental information into the investment business. It *includes environmental indicators* of a general nature, such as energy consumption, emission figures and waste quantities, and for specific industries. The defined financial indicators include running environmental expenses for quality assurance and outlays on energy and disposal, as well as environment-related investments and provisions and the associated depreciation. The *qualitative criteria* include statements on the company's strategy for the environment, environmental management, and process and product strategies with concrete targets.

This uniform format will enable analysts to add environmental criteria to their valuation process. An interesting *strategic component* is how the company might be able to create value with its commitment to the environment, and whether it is in a position to anticipate the future influence of lawmakers and markets. Environmental factors such as penalties for liability, compliance with the law and risk management can be integrated into the existing analytical framework under the heading of *risk aspects*. Companies can also achieve *competitive advantages* with innovative products and services as well as eco efficient process controls and a reduction in environment-related costs. Thus the analysts' recommendations will be expanded to include an environmental component to help investors make their investment decisions.

Implementing the Concept of "Environmental Shareholder Value"

Environmental Rating Agencies

Various research and documentation centres have sprung up since the '70s to compile company information. The Investor Responsibility Research Center (IRRC, USA) was established at the height of the anti-Vietnam War movement, when pension funds came under increasing pressure to strip defence industry shares from their portfolios. Among the products offered by the IRRC is a corporate environmental profile, in which the biggest US companies are ranked for their environmental performance and misconduct (e.g. penalties) and compared against industry benchmarks. In the United Kingdom, the Ethical Investment Research Service (EIRIS) was founded by a number of organizations, primarily religious groups, to focus above all on companies' ethical behaviour. In the German-speaking region, institutions such as FIFEGA (Forschungsinstitut für ethisch-ökologische Geldanlagen), Ökom, the Centre Info in Fribourg (Switzerland) and the Institut fur Ökonomie und Ökumene Südwind offer information materials based on ethical and ecological criteria to help investors with their decisions. As is the case with environmental reporting, however, the external environmental rating organizations also have no uniform standards. They have therefore developed their own catalogues of assessment criteria to meet their own needs.

Ecology, Financial Analysis and the Investment Business

Although the opinions of financial analysts vary as to the relevance of environmental criteria, investment products have already been specially created to embody an environmental dimension. In the English-speaking world (USA, UK) green funds are primarily sold as ethical investments, or as screening out specific sectors such as alcoholic beverages, the weapons industry and/or nuclear power. The market volume is relatively large; at the end of June 1998, USD 18.5 billion were invested in ethical/ecological funds in the USA and some DEM 4 billion in the UK (Öko-Invest 166/98). In the German-speaking market the funds on offer are dominated by companies involved in environmental technology (recycling firms, suppliers of filtration plant and equipment, renewable energy sources). At DEM 980 million the market volume is relatively small, because the lack of scope for diversification means that satisfactory performance is relatively infrequent. Various funds have already been dissolved because of the modest volumes. Others, such as Oeco-Protec from Credit Suisse, have been refocused around the concept of eco-efficiency, which is also the guiding principle of more recent funds such as Oeco Sar, launched by Bank Sarasin of Basel in 1994, and the Norwegian Storebrand-Scudder Environmental Value Fund. These funds follow a "best-in-class" approach, investing in the environmental leaders in each sector. Depending on the philosophy, companies that generate a large percentage of their total sales from certain activities may be blacklisted.

Environmental Performance Analysis at Swiss Bank Corporation (now UBS)

Encouraged by the growing customer demand for investment vehicles with an environmental angle and by a conviction that consideration of ecological factors improves the quality of information produced by financial analysis, an "Environmental Performance Analysis" group was set up in SBC's Institutional Asset Management business area in 1996. Its task is to develop and apply a methodology for the evaluation of companies' environmental performance. The process draws on outside experts as well as employing the now established tools of eco-ratings and environmental accounting. Active participation in various working groups such as the WBCSD task force mentioned above also provides valuable feedback and ideas, as do the direct contacts with industry trade groups. It is particularly important to maintain close contact with a "scientific expert board" which includes two co-authors of *Factor Four*, Amory Lovins and Ernst Ulrich von Weizsäcker, as well as Swiss professor Ruth Kaufmann. The message in *Factor Four*, that economic and ecological performance can be optimized and prosperity can be doubled while resource consumption is cut in half, provided that resources are used more efficiently, is one of the scientific foundations of the bank's approach to environmental issues, together with the WBCSD's concept of eco-efficiency.

Both concepts are aimed at reducing the consumption of energy and resources per unit of added value or utility. To put it another way: more value is created with a lower burden on the environment. Eco-efficient companies not only minimize their environmental risks but also generate competitive advantages with their ecologically-driven activities.

Pursuing this line of thought, the analytical method developed by UBS identifies the companies in each sector that recognize environmental aspects as a strategic challenge and long-range opportunity. This approach is intended to support the integration of environmental responsibility into general economic activities across the board, rather than encouraging the development of niche markets.

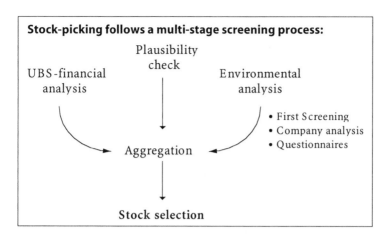

Fig. 4.   Selection process

I. *Environmental policy and strategy*	Corporate guidelines and long term environmental strategy
II. *Environmental management*	Environmental management systems and programs, controlling, auditing, internal measures including training, risk management
III. *Ecology-related costs & savings*	Scale of costs and savings, integration into cost accounting
IV. *Environmental communications*	Information to interested parties outside the company
V. *Process strategies*	Concrete objectives and measures, environmental audits
VI. *Environmental statistics (input/output)*	Energy & water consumption, emissions & waste
VII. *Environmental product strategies*	Purchasing guidelines, eco-efficiency of products, life cycle assessment, customer service, return/take-back & recycling systems

Fig. 5.   Environmental performance ratings at UBS

The related Eco Performance portfolios invest exclusively in companies with an excellent track record of environmental and economic performance. The innovative element is the twofold strategy: the funds consider companies that manifest an above-average commitment to environmental protection (the so-called eco-leaders) as well as companies whose products embody a high degree of resource efficiency (the so-called eco-innovators).

The *eco-leaders* are firms or corporate groups that control their impact on the environment with the help of environmental management systems while successively boosting their eco-efficiency. The *eco-innovators* are companies whose products and services contribute to the achievement of a particular economic benefit or utility with the highest possible degree of resource efficiency. These firms are selected mainly on the basis of their product range (examples: renewable sources of energy, alternative mobility concepts).

1. Financial analysis rating based on traditional criteria.
2. Environmental analysis and rating. The eco-efficiency of products and processes is evaluated using a method developed by UBS. The research is based on the company's own data sources, including environmental performance reports, and on the results of direct surveys using a comprehensive questionnaire. The questionnaire covers seven main areas with the following analytical criteria:

    This analytical framework is adjusted for specific industries to identify and evaluate the strategies worked out in each case for the ecological challenges they face.

    • In the automotive industry, the targeted criteria include (among others) reductions in fuel consumption across the entire vehicle fleet and the development of alternative concepts of mobility.
    • For electronics companies, considerable weight is given to strategies to reduce energy consumption and increase product life, as well as return policies/take-back systems and recycling systems.
    • Wholesalers and retailers are judged on their ecological purchasing guidelines, the provision of environmental information to consumers and transportation management, among other points.
    • Energy utilities can score well with renewable energy sources, customer counseling and financial incentives to conserve power as well as net metering offers.
    • Banks are scrutinised for measures to reduce consumption of energy and paper and the integration of ecological risks into their loan approval process, plus the range of environmental investment products offered to customers.

3. All this information is supplemented by input from the daily press, specialist journals, databases and electronic media such as the Internet. A contract has been signed with an outside environmental media office to gather documentation for regular plausibility checks of the ratings.
4. The final step in admission to the Eco Performance portfolios is an aggregation of the financial and ecological ratings. The portfolios only invest in firms that score well on both counts.

The goal of this investment policy is to achieve the highest possible capital growth over the long term with due consideration of ecological criteria. This involves a synthesis of economic and ecological assessments (ecology is economy with a future), offering the investor an optimal combination of superior environmental benefits and a return in line with the market.

This target is followed by an industry and country weighting based on the MSCI World. Comparing the performance of the Eco Performance portfolio with that of the MSCI Index World over the last three years confirms the validity of this concept: the MSCI was clearly outperformed. The result of this backtest is also endorsed with the performance of the portfolios since their public launch in June 1997. The volume of more than 200 mio CHF after one year shows in the fund for private investors and a corresponding one for institutional investors shows that investors can be attracted to this concept.

Developments in the Swiss market show that many of the suppliers of investment products, at least, are convinced by this new perspective. The participating big banks and an environmental money management firm have selected this approach to offer their customers attractive products with a market return. This new kind of environmental commitment is going to shift the parameters away from the hitherto prevalent niche-oriented strategies.

These banks believe not only in the ecological utility of their products but also in the economic potential of their method. The targeted customer segment is equally interested in economic *and* environmental performance. The message to these investors runs: if you want to invest in the future, put your money in environmentally efficient companies. This is the message that can convince the financial markets of the value of ecological information and environmental investment.

UBS (Founded June 29, 1998 by a Merger of Swiss Bank Corporation and Union Bank of Switzerland)

UBS is a "universal" or full-services bank with some 55 000 employees around the world, organized into five core businesses: UBS Private and Corporate Clients, Warburg Dillon Read (Investment Banking), UBS Brinson (Institutional Asset Management), UBS Private Banking and Private Equity Business.

Both banks, Swiss Bank Corporation and Union Bank of Switzerland have had an environmental policy since the early 90ies, incorporating concrete objectives for efficient use of resources and a commitment to the consideration of environmental factors in its products and services as well as in its training and information activities. Both banks have signed the UNEP declaration, "Banking and the Environment, a Statement by Banks on the Environment and Sustainable Development".

The environmental management system now includes a co-ordinating department within the Corporate Risk Management Department focusing on Investment Banking and various specialized units in the areas of Operational Ecology, Human Resources and Training and Credit Risk Management. In order to integrate ecological criteria into the xx, a task force in the institutional asset management division developed a methodology for the application of environmental performance analysis to companies. Building on this, the "Environmental Performance – World Equities" investment group of the AST Investment Foundation for Swiss employee benefit plans was set up in 1997, together with an investment fund under Luxembourg law. The UBS task force also handles individual assignments for institutional customers.

References

ACBE, Advisory Committee on Business and the Environment (1997): Environmental Reporting and the Financial Sector: Draft Guidelines on Good Practises, A Consultation Paper. London.

Deml, Baumgarten, Bobikiewicz (1994): *Grünes Geld, Jahrbuch für ethisch-ökologische Geldanlagen 1995/96*, 2nd edition. Service Fachverlag, Vienna.

Hassler, R. (1994): Öko-Rating – *Ökologische Unternehmensbewertung als neues Informationsinstrument*, Ökom-Schriftenreihe zur ökologischen Kommunikation, Munich.

Idee V (Gesellschaft fur ökologische Dienstleistungen mbH) (1994): *Ökologisches Investment. Marktübersicht über ökologisches Investment.* Heilbronn.

Knörzer, A. (1996): *Ökologische Aspekte im Investment Research*, Swiss Banking School Publication 137, Paul Haupt, Bern.

Müller; de Frutos; Schüssler; Haarbosch (1996): *Eco-Efficiency and financial analysis. The financial analyst's view.* The European Federation of Financial Analysts Societies.

n.n. (1994): City Analysts and the Environment: A survey of environmental attitudes in the City of London. Business in the Environment/Extel Financial Ltd., London.

Öko-Invest No. 166/98, pp. 6f., Vienna, July 1998.

Overlack-Kosel et al. (1995): *Kreditrisiken aus Umweltrisiken.* Economica Verlag, Bonn.

Schmidheiny, Zorraquin, with the WBCSD (1996): Finanzierung des Kurswechsels. Die Finanzmärkte als Schrittmacher der Ökoeffizienz. Best Business Books AG, Zurich.

VfU (publisher) (1996): Umweltberichterstattung von Finanzdienstleistern. Bad Honnef.

Weizsäcker; Lovins; Lovins (1996): *Factor Four – Doubling Wealth, Halving Resource Use.* Report to the Club of Rome, Droemer Knaur, 9th edition, revised, Munich.

World Business Council for Sustainable Development (1997): *Environmental Performance and Shareholder Value.* Geneva.

Empirical Studies on the Relationship Between Environmental and Economic Performance

Cohen, M. (1995): *Environmental and Financial Performance: Are they related?* in: Environmental Information Service of the Investor Responsibility Research Center, Washington, D.C.

Hamilton, J. (1993): Pollution as news: media and stock market reactions to the toxic release inventory data. Duke University.

Hart, S.H.; Ahuja, G. (1995): Does it Pay to be Green? An Empirical Examination of the Relationship between Pollution Prevention and Firm Performance. University of Michigan.

Johnson, S.D. (1995): An Analysis of the relationship between Corporate Environmental and Economic Performance at the level of the Firm. University of California.

White, M. (1995): *Corporate Environmental Performance and Shareholder Value.* Mclntre School of Commerce at the University of Virginia.

White, M. (1995): *Investor Response to the Exxon Valdez Oil Spill.* Mclntre School of Commerce at the University of Virginia.

B

Best Practice Approaches in Environmental Management

1 New Concepts in Environmental Auditing – The Application of Auditing Techniques to Environmental Management Systems

Martin Houldin

Introduction

The introduction of European and international standards and schemes for environmental management sytems (in particular the Eco Management and Audit Scheme – EMAS – and ISO 14001) has led to the latest development in a long history of environmental auditing. Perhaps of greatest significance is the fact that this development is introducing new qualifications and professionals into the field of environmental auditing. In many cases the new breed of 'environmental auditors' do not have either a technical environmental background or any substantial environmental management experience. What are the implications of these developments? Are we diluting 'environmental expertise' with this rapid growth in environment management systems (EMS) auditing, and the introduction of professionals from other fields of auditing? Is this good or bad news for the value and credibility of environmental management systems?

In this article I hope to show that EMS auditing has so far been developed on reasonably sound principles. What is most important, however, is that we continue to follow these principles as growth in EMS auditing continues, and that consequently the contribution made by auditing to the effectiveness of environmental management systems may be duly recognised. Essentially I believe that the credibility of EMS standards rests to a great extent on the strength of EMS auditor qualifications and auditing techniques. The value of EMS certificates / registrations will stand or fall on the quality of EMS auditing.

Definitions and Types of Environmental Auditing: The Need for new Definitions and Clearer Differentiation Between Audit Objectives

There is little consensus on the definition of environmental auditing. Anybody who regularly attends environmental conferences will hear various references; in some cases six types of environmental auditing, or 10 types or 20 types.

It is not for me to attempt any 'new' definition. However, we may see things a little more clearly if we make reference to other existing definitions, and to the differences in scope and objectives of auditing as a management tool.

Other auditing professions, such as financial and internal auditors, use a stricter definition than is typically the case in environmental auditing. An audit can be defined as 'the verification of a statement of compliance or conformity with defined requirements and conditions, made by the auditee either explicitly or implicitly'. For example, in the case of financial audits of company accounts, the auditee (the company) presents its statement of accounts as giving a 'true and fair view' of its financial performance in the recent period, and of its current financial position. The external financial auditor seeks to verify this explicit statement, and that the implicit part (that there is conformity with accounting standards) can also be verified. Another example might be where a team of internal auditors are carrying out an operational audit. In this case, the auditee (the division or department) makes an implicit statement that it complies with all relevant company policies and procedures. The auditor seeks to verify this implicit statement.

If we attempt to apply this more traditional and established definition of auditing – which always includes some reference to 'defined requirements and conditions' – to existing environmental auditing definitions and practice, we see that some types of environmental audit are not strictly auditing. In some cases the objectives are different from the verification of a statement (implicit or explicit) or specified conditions.

One of the more recognised definitions of environmental auditing, which provides a better fit to the more general practice in auditing, is that provided by the International Chamber of Commerce (ICC) in 1989. "A management tool comprising a systematic, documented, periodic and objective evaluation of how well environmental organisation, management and equipment are performing with the aim of helping to safeguard the environment by:

a facilitating management control of environmental practices;
b assessing compliance with company policies, which would include meeting regulatory requirements"

Objectives in Auditng

Reference to a stricter definition of auditing, and to the different objectives management has in commissioning audits, enables us to recognise two broad categories of environmental audit.

First, there are audits where the objectives are primarily to gather information. Commonly used terms are 'environmental review', 'environmental issues audit', 'site audit'. In these cases management requires information as a basis for determining policy and decision-making, whether strategic (in relation to business and environmental issues) or operational (in relation to identified environmental problems).

Table 1.   Objectives in environmental auditing

SCOPE/OBJECTIVES	INFORMATION	VERIFICATION
Regulatory Compliance		Legal requirements
Due Diligence	Potential liability/cost Asset valuation	Legal requirements
Impact Assessment	Actual/potential impact Presentation to public	Compliance with planning law
Management Control (internal audit)	Effectiveness of training Performance vs. Objectives Programme progress	Compliance with policy, procedures and internal systems
EMS Certification		Conformance with specified requirements (ISO 14001/EMAS)
Environmental Reporting	Environmental issues, performance and related costs for publication	Validation of statement for completeness and reliability
Management Review	Issues and aspects for decision- making Scope for improvement	

Secondly, there are audits where the objectives are to verify a 'statement of compliance or conformity with defined requirement and conditions'. Examples include 'regulatory compliance audits', 'EMAS statement validation', as well as EMS audits (whether internal or external).

Table 1 identifies the typical information gathering and verification objectives in different types of 'audits'.

Until perhaps 1993/94 most environmental audits were of an information gathering nature, as a tool for management decision-making. The exceptions to this were the audits carried out to verify legal and regulatory compliance against specific conditions and requirements. The introduction of EMS standards, as well as a growing trend towards the publication of environmental reports, has led to a significant increase in 'verification' audits.

A Brief History of Environmental Auditing and the Introduction of EMS Audits

There are clearly different ways of looking at the history of developments in environmental auditing, and much has been written about auditing over the years. As a background for the latest developments it is useful to refer to three major phases.

Phase I has its roots in the 1970s with the key developments in environmental legislation in the USA, particularly related to impact assessments (NEPA), liabilities for contamination (CERCLA) and the acts for clean air and pollution prevention. The focus of environmental auditing, therefore, through the 1970s and a large part of the 1980s was linked to legislative and regulatory compliance, and to assessments of liabilities and impacts. The main objective was the avoidance of penalties and/or liabilities.

Phase II I believe became evident in the mid-1980s, mainly in Europe, although this does not mean that there was no similar practice earlier or elsewhere. In this phase the focus of the audits became, additionally, to identify environmental issues related to markets as well as legislation. A number of factors, during the mid to late 1980s had prompted a number of companies (not only heavy polluters) to recognise environmental issues as business issues of strategic (as well as operational) importance. It is difficult to summarise a large number of important developments over a period of more than five years. However, I believe that there were two key factors.

Firstly, that European Community (as it was then) environmental policy shifted more towards the inclusion of market and economic instruments to be used alongside traditional command and control legislation. This introduced the concept of the use of 'voluntary' industry initiatives and market pressures, as the alternative to controls, being more economic, and possibly more effective, in terms of environmental control and pollution prevention.

Secondly, there was a significant shift in consumer and political interests in environmental issues, associated not only with pollution control, but also wider issues related to products – use of non-renewable resources and the impacts associated with customer use and final disposal. It became fashionable to talk of a variety of stakeholders and their interests.

Over a relatively short period the management of environmental issues came to involve not only technical or engineering departments, but almost every other function in the organisation, associated with product design, sales and marketing, purchasing and even finance. Companies began to develop much broader policies and strategies for the environment; some began to produce reports on environmental issues and performance, as a means of demonstrating the seriousness with which they were taken.

Phase III in the early 1990s, saw the development and introduction of a number of standards and schemes for environmental management. The European Commission developed the Eco-Audit (later to become the Eco-Management and Audit Scheme – EMAS) regulation. At the same time, and very much in tandem, the British Standards Institute developed BS 7750, a standard for environmental management systems. Soon after other countries developed their own national standards, or decided to adopt BS 7750. CEN – the European standards organisation – was mandated by the European Commission to develop a European standard, or recommend recognition of existing standards. By 1992/93, the International Standards Organisation (ISO) had established an advisory group (SAGE) for the development of an ISO standard for environmental management systems. By 1996, in record time, we had ISO 14401 as well as EMAS (finally introduced in Member States in 1995). By the end of 1997 there will be no EMS certifications to national standards, and we expect that ISO 14001 will be fully recognised by the European Commission as meeting the corresponding environmental management systems requirements.

However, the introduction of these standards does not represent the only significant environmental management development. At the industry sector level perhaps the most important development has been the extension of the Chemical Industries 'Responsible Care' programme, from health and safety into health, safety and environment (HSE). While the scheme is not used as a normative standard in itself, it has been used by some as a means of integrating HSE with quality management systems in ISO 9000.

Some major international companies have resisted the implementation of formal, structured standards for environmental management systems, and have developed their own approaches to environmental control and improvement. Notably, some of these focus on the application of life-cycle analysis techniques as a means of identifying opportunities for improvement.

The remainder of this paper focuses on the standards for EMS, and in particular their links with environmental auditing.

Auditing in EMS

As the majority of these standards have been developed alongside each other it is perhaps not surprising that there is little difference between them. There is in fact only one substantial difference between EMAS and ISO 14001, in that EMAS re-

Table 2. Auditing requirements related to ISO 14001 and EMAS

	Internal	External
EMAS	Article 3 (d)	Article 3 (g)
	Article 4	Article 4
	Annex I B (6)	Annex III
	Annex II	
ISO 14001	Clause 4.5.4	• Voluntary/market demand • Certification to follow accreditation criteria and guidelines (EAC/G5)

quires the publication of an audited statement of environmental performance, in addition to the common management systems requirements. This does have some significant implications for environmental auditing.

There are, however, two different aspects of environmental auditing which we must consider. First, internal environmental auditing, carried out either by first or second parties. Second, external auditing, carried out by independent third parties.

Internal Environmental Auditing

Both EMAS and ISO 14001 require internal auditing. Although EMAS is more specific and detailed in its requirements (referring to ISO 14012) in essence there is no difference in how internal audits are planned and conducted. Many companies, of course, have had the similar experience of internal auditing for ISO 9000.

The focus of the internal environmental management audit is to check whether the company's own systems and procedures are being adhered to, and whether they are effective in helping to achieve environmental objectives and targets, in line with the environmental policy. The key elements of the auditing process, therefore, will normally be:

- A team of suitably trained and qualified internal auditors,
- A schedule for auditing different components of the EMS as relevant to each organisational unit, identifying those which may be important enough to audit more frequently,
- Audit checklists covering the auditable requirements of the EMS, as they apply to organisational units,
- Audit reports, including non-compliances raised.

Those who are familiar with the requirements of ISO 1012 and ISO 14012 (Guidelines for Environmental Auditing – Qualification Criteria for Environmental Auditors) will understand that the basic auditing process follows recognised good practice in internal auditing, whether it be for quality management systems, EMS or other operational/financial audits.

The main issue, therefore, for EMS internal auditing is the extent of environmentally related qualifications and/or training required.

External Environmental Management Auditing

In principle and in practice both EMAS and ISO 14001 involve external auditing by independent third party auditors. Technically, however, there is a difference that can be important.

EMAS specifies the requirement for external audit (verification) in the Regulation document itself; whereas the ISO 14001 standard, as with other similar standards, makes no requirement for external audit. It is merely stated that the standard is a normative document that may be used for 'self-assessment' or independent third party auditing purposes.

It is left, therefore, to other documents to specify the requirements for external auditing. At the European, and increasingly at the international levels, this is done through the accreditation systems in each country. Accreditation is in effect a form of certification, except that it is the 'certification' of the third party auditors (or certification bodies). Certification Bodies are accredited by Accreditation Bodies to carry out EMS (or other) certification audits. In true auditing style, however, this must be done in accordance with defined criteria, which are now provided at the European level, and soon will be at the international level.

In Europe, the accreditation criteria have been set by the EAC (European Accreditation of Certification) in document "EAC – G5 Guidelines for the Accreditation of Certification Bodies for Environmental Management Systems". A similar document, based on the European criteria, is currently being developed by the International Accreditation Forum (IAF) – "General Requirements for Bodies Operating Assessment and Certification/Registration of Environmental Management Systems (EMS)".

There are naturally differences in approach and interpretation of the standards between different parts of the world. However, there is general agreement that the credibility of EMS certification will depend largely on the quality of auditing. But, how do we define 'quality' in auditing? The words 'rigorous' and 'comprehensive' perhaps spring to mind. Of even greater importance, in my view as business becomes more international, is the consistency of auditing. Consistency is important not just between audits in one country, but even more so between audits in different parts of the world. The key factors that influence consistency must therefore be determined. These include the methods and requirements for objective evidence applied to auditing, and the competence (or qualifications) of the auditors themselves. In the case of EMS auditing environmental (in addition to audit skills and sectoral) competence is a key factor, even more so because ISO 14001 may be described as being performance oriented. Absolute standards of performance are *not* set, but companies must set their own objectives and targets, and be able to demonstrate control and improvement in environmental performance (environmental effects/aspects).

The guidelines for accreditation, although there are a number of clauses, cover the two major areas of audit methodology and auditor (or audit team) competence.

Auditor Qualifications/Competence

The wider introduction of environmental management systems auditing has a number of important implications for companies undertaking certification, and certification bodies alike.

Internal Auditing

For many companies, environmental auditing to date has involved environmental reviews and / or audits of technical and legal specifications. EMS auditing therefore introduces an additional resource need, which companies very often have to satisfy within existing staffing budgets. However, a significant number of companies in many parts of Europe have achieved ISO 9001/2 certification, and therefore have experience of internal auditing of quality management systems. As the basic process is or can be common between quality and environmental management systems audits, many companies include existing quality auditors on the EMS auditing teams. Typically they are joined by environmental managers and environmental engineering staff.

This introduces different training needs for those EMS auditing team members with different backgrounds. Some will not need auditing skills training but will need environmental and EMS training. Others will need auditing skills rather than environmental training. The extent to which internal auditing team members are able, or should be required, to meet the specific educational requirements of ISO 14012 (Auditor Qualifications) is a question that is often raised. Other parts of the guidelines refer to 'relevant / appropriate work experience and training'. In practice, auditors will be people who have long experience and in-depth knowledge of the company, its organisation and its processes, in addition to auditing and environmental training totalling between one and three weeks.

External Auditors

The external auditor has a different objective and a different kind of responsibility. To a greater or lesser extent there will be people – or stakeholders – who will seek to gain some assurance from the company's or site's certification to the EMS standard. The quality of the audit, therefore, has a major role to play in ensuring the credibility and value of certification, and the reliance that stakeholders can place in the certification process.

The auditors' objective is to check that the requirements of the relevant standard are met, and that the systems are effective in demonstrating regulatory compliance and continual improvement. For this reason the accreditation criteria guide certification bodies to determine competence in four key areas:

- Environmental aspects and relevant management practices,
- Management systems auditing,
- Relevant processes and activities (organisational, business and industrial),
- The relevant standard(s).

For me, the real test of an auditor of management systems, which have defined performance objectives, is whether he or she is adequately qualified to detect errors of omission. In other words, what are the environmental effects / aspects not covered; which records are being inadequately reported; how should performance be monitored (data and information management)? Many would agree that it is relatively easy to audit systems as they are presented to the auditor; but the ability to identify inadequacies in the extent of environmental management is a good indication of competence. The auditor, using available information related to activities, should be able to determine what he or she expects to find. This means that the more highly qualified auditors will have knowledge and experience of:

- A wide range of effects/aspects in relation to activities and processes,
- Recognised environmental management practices,
- Techniques for measurement and quantification of environmental performance, in particular what could reasonably be expected given the nature and scale of activities and processes.

Certification bodies are required to demonstrate that they have sufficient competence at management levels to assess auditor qualifications, and to determine the team competence required for a particular assessment.

The difficulty lies in being able to determine the 'appropriate' level of experience, training and qualification. ISO 14012 does not give any specific guidelines to help certification bodies, referring only to the need for 'appropriate and relevant work experience and training'.

In the absence of more specific guidance, there is a risk that the quality of auditing may suffer because of inadequately qualified auditors. There are those – from the environmental community and quality management auditing fields – who believe that a short training course (often as short as 2–5 days) is adequate. While it is recognised that such training can qualify people to participate on EMS auditing teams, there is some risk that commercial pressures will lead to short cuts. The classic errors are, firstly, that some consultants have a tendency to review and gather information rather than verify or audit and, secondly, that auditors from other fields do not have adequate environmental management competence to recognise what may be an ineffective system, or to detect errors of omission.

Work is, however, currently under way to produce further guidance that will provide a structured approach to determining environmental competence. It is likely that three levels of competence – general, specific and expert – will be recognised, in relation to different types of activities. Also, guidance will cover the need to

Table 3. Example competence requirements (illustration purposes only)

Automotive Manufacturer	Air	Land	Water	Natural Resource
General				
Specialist				
Expert				

determine the appropriate level of competence in each environmental area (air, land, water, natural resources). For different types of industrial and business activities, general guidance will be given as to the appropriate level of experience in each area.

Environmental Verifiers

The environmental verifier has a specific role, introduced by EMAS. In addition to auditing the EMS, the verifier must audit or validate the environmental statement (to be published). This process requires verifiers to:

- Identify and check procedures and controls for data gathering,
- Follow the audit trail for specific data reported and claims made in the statement,
- Assess the statement within its environmental context,
- Assess the statement's language, style and presentation for the adequacy of its likely interpretation and comprehension by stakeholders.

This process is different to such an extent that the verifier is required to have a much higher level of environmental competence, compared with EMS auditors, and needs to be familiar with data and information management processes. While the statement validation can be, and often is, integrated with the EMS audit, people should not interpret this as meaning that somebody competent to carry out the EMS audit is necessarily (automatically) adequately qualified to validate the environmental statement.

Key Issues

The introduction of EMS auditing, largely through the development of EMS standards (such as ISO 14001 and EMAS) raises some important issues which must be resolved within the next one to two years. Otherwise there is some risk that credibility of EMS certification will not meet the expectations raised with a variety of stakeholders and interested parties.

These issues can be summarised as follows:

- Are companies suffering from audit overload? If we accept the need for EMS internal auditing how can this be met, with reasonably and adequately competent auditors, without breaking the bank? More specifically, how should internal auditors interpret the requirements of ISO 14012?
- How do we encourage adequate levels of qualification and competence among certification bodies world-wide, to have a good chance of achieving interna-

tional consistency and credibility in auditing quality? At present I believe that the sudden influx of 'new' environmental auditors represents a considerable risk. Again it would seem that a reasonable and sound, but appropriate interpretation must be made of ISO 14012 (different from that for internal auditors).

I do not pretend that these are the only important issues of the day in EMS auditing, or more generally in environmental auditing. However, if we can resolve the issues of suitable environmental qualifications and competence for internal and external auditors, in a way that is both economic for companies and certification bodies, and environmentally sound, then we will have taken a big step towards a better environment for the next century.

2 Waste Is Good Material at the Wrong Place – Perspectives of a Recycling Economy

José A. Lutzenberger

When I lived in Brasilia, I possessed a beautiful, precious collection of orchids on my porch. I am no collector of orchids, but I knew a florist's shop in the city center that received blossoming plants from orchid cultivators of São Paolo, which were then sold at the fourfold price. Plants beginning to fade and hence not meeting an immediate buyer were sent to the landfill, among them very precious specimens, up to 15 years old, which clearly had not been grown in a conservatory but had been stolen from nature. I cannot stand such things; I fetched them. They rewarded me in the coming year with the renewed magnificent beauty of their flowers. For the florist, these wonderful flowers were a means to get money, nothing else. We find this mentality all across our modern industrial societies. It is even dominating a major part of waste management.

At last, however, the powerful technocrats and bureaucrats have been reached by the insight that we cannot go on for long in such a way. The Rio Summit of 1992, which I could help to organize, focussed on 'Sustainable Development'. If we do not want to go on stealing the future from our offspring, we have to fundamentally change our way of thinking, and not least in the field of waste management. Even "orderly" landfills or incineration are false solutions, they are the contrary of being sustainable.

All our modern technologies are based on certain, often implicit ideologies; they rely on certain postulates which are nothing but beliefs, yet are meant to be understood as common sense.

Let us look at the basic postulate of waste management. Waste management starts from the assumption that there are inherently bad, worthless, harmful things – we call them junk, waste, rubbish etc. and simply want to get rid of them, like those tanners who simply throw their leather scraps onto the street, or like the mentioned florist.

Yet we cannot get rid of anything on this planet, nothing can disappear. Fortunately it is still too expensive to launch things into space. And actually there is nothing bad 'per se'. A good glass of good wine on the dinner table is great. If I spill it on the carpet, it is still wine, but it has become filth. Junk is always good material at the wrong place.

Even radioactive stuff is nothing bad if located in its proper place, in the sun, or in the stars, or in the Earth's inner depths. It even has life-preserving functions then. But it has nothing to do in the biosphere, as wine has nothing to do on the carpet. Therefore, we must not operate any nuclear plants, since otherwise we cannot avoid to have the life-preserving systems of this living planet poisoned for millennia.

Second postulate of modern waste management: solutions to any problem should be as centralistic and large-scale as possible. Some years ago, there was a project to build one single enormous waste water treatment facility for São Paolo, designed for the twenty millions of inhabitants forecasted for the year 2000. Imagine the torn-up streets, the gigantic roadworks for installing a sewage system, the many pumping stations ... In some directions, São Paolo extends today to a length of 100 km. Fortunately, there was not enough money for this!

In 1992 and 1993, I visited two major fairs for 'environmental technology', the UTEC-Absorga in Vienna and the IFAT in Munich. What you find at such fairs is almost exclusively highly complex, extremely expensive and centralistically oriented technology, for instance such absurd facilities as plants for drying and burning sewage sludge. Nearly the only offer to be called an ecological solution was a cheap waste water treatment plant for urban sewage, presented by an American firm. It operated with swimming water plants in simple ponds. Other offers went as far as suggesting to put out of service dozens of existing small sewage works at village level and to replace them by a large-scale, central facility with a huge net of sewerage and many pumping stations. Of course, such proposals have their inherent logic: expensive projects, millions of Deutschmarks for papers only, billions for the construction business if possible ... what a pleasant perspective for those who profit from this corruption!

In addition to resulting in high costs, this approach yields "solutions" which almost inevitably are unecological, socially undesirable, and moreover not designed to be pursued for long, above all not in the highly-industrialized and densely populated nations of the so-called first world. All of you should once take the time to stroll around the large "orderly" landfills in Europe and the USA as well as the disorderly or illicit landfills in Brazil! When I was a member of the Brazilian government, the state government of Pernambuco received the offer of an enormous waste incineration plant for the town of Recife, to be constructed gratuitously. The snag was: this facility was designed to accept an additional 2.7 millions of tons of European industrial waste. We came just in time to prevent this.

The fact that the majority of Brazilian towns have not yet any sewage works, nor sensible solutions for the waste issue, the fact that so many industries make hardly any efforts to solve their pollution problems, might be mainly due to that approach: town mayors and plant directors are faced with only just this kind of expensive solutions. Most mayors, it is true, would be happy with expensive projects, but fortunately they lack the means. Business people, on the other hand, who calculate very tightly, refuse such solutions as long as possible. The Environmental Agency is not much of a help, either. At first, it may take months for a project to be submitted, then the authority needs even more time to license or reject it, then follows another extended period until a revised project is submitted etc. Meanwhile, nothing happens. Finally, the expensive projects often work badly, or not at all, and the whole procedure starts again.

For more than 15 years already, Brazilian tanneries are urged to instal waste water treatment facilities. Meanwhile, most of them have basins for primary sedimentation and basins for ventilation. But many – as far as I know, the majority – still discharge the chrome used in tanning into their waste water, they do not separate their lime bath from other waste water either, and consequently they are looking for approriate landfills to deposit their highly chrome-contaminated residual sludges. The situation is even more severe with galvanization facilities. There, the dangerous chromium IV and other heavy metals go into the waste water.

The paradigm of predominant waste management has to be completely reversed. The first question will have to be "What can we make out of it?" instead of "How do we get rid of it?". Second, we have to search for holistically designed, stepwise proceeding, evolutional-systemic solutions. We, that is my waste processing firm VIDA Produtos Biológicos Ltda., are today able to recycle the solid wastes of a large pulp production plant (RIOCELL), amounting to 700 tons a day: residual sludges, lime sludges, hard coal ash, barks, wood dust, waste, scrap and debris. This saves the plant annually more than half a million dollars. We live on the returns achieved from the valuable materials. A conventional solution would have cost millions, and available space for landfills would already have been exhausted by now. We proceeded step by step. One partial solution induced the subsequent one. Our mistakes were small and cheap ones. Had we proceeded according to routines provided by the Environmental Agency, we could not have started at all, since we were not able to submit a complete detailed project, nor did we want to.

If I proceed systemically in the tannery, I first focus on the lime and the chrome baths. If those two waste waters do no longer come together, the stench also ceases. With very low investments, which pay in very few time, it is possible to precipitate the chrome and recycle it into the process. As to the lime, the bath can be reused time and again, after sedimentation of the sludge, titration and supplementing. Some tanners even assured me that this had improved leather quality, apart from saving several additives. The sedimented sludge may be spread straight on farmland, this will yield maize plants of three meters height. The sulphide will oxidize to sulphate – Brazilian soils are almost everywhere poor in sulphur – and lime is good for the soil anyway. Another possibility would be to use the sludge – after adjusting to low-alkaline pH-value – via biogas. This would provide the farm with some energy, and the mature liquid manure will then be even more beneficial for agriculture.

Given the simplicity of precipitation and recycling of chrome in a tannery, I wonder why not yet all tanneries employ that procedure. The precipitation basins both for the lime liquor and for chrome waste water are that simple and cheap that they will pay in a few months already.

Given sensibly designed waste water pipes within a tannery and given clean working, waste water can then be kept sufficiently free from chrome, so that the resulting sludge has not necessarily to be deposited in landfills for hazardous wastes. An alternative would be to spread it in eucalyptus woods or in acacia plantations. Brazil's best soils are those originating from the basalt of the Serra Geral. This rock has up to more than 200 ppm of chrome. I do not mean to disregard dangers from heavy metal contamination. Yet nevertheless, some differentiation is needed. Cadmium, mercury and lead, as well as arsenic and antimony require much attention, and so do radioactive actinoides. In my opinion, zinc, copper, manganese, molybdenum or cobalt cause only minor problems. As long as concentrations remain low, these elements count among the necessary trace elements for living beings. Chrome is another trace element indispensable for life.

I have been engaged in waste management issues for more than twenty years, although actually I am an agronomist. I do so because I want to foster sound agriculture. Let's face the facts: today there are hundreds of millions of hectares of finest soils, particularly in Brazil, which are systematically degraded or finally destroyed by loss of humus, poisoning of soil life, erosion, unecological agricultural methods etc. At the same time, hundreds of millions of tons of industrial and urban wastes are produced, which consist of valuable organic substances, but which nevertheless are understood only as waste or as waste water, and which consequently are "disposed", mostly by means of procedures resulting in total loss or irreversible contamination. There are even laws, of course suggested by certain lobbies, which prohibit a good many reasonable recycling methods.

The tannery 'Ritter' provides a fine example of recycling. Here, pig food is prepared from meat remainders and scraps of raw hides. The liquid manure from the pigs then goes to a large apple tree plantation and to maize fields. The pig food is composed of animal proteins, cooked meat remainders, and leather scraps, mixed with maize flour. About 12 000 pigs are raised on this diet every year.

Almost no pesticides are used, neither for the apples nor for the maize. The model is blemished only by the fact that the pigs are kept in conventional 'prisons', without any run. There is hardly one square meter per animal in porker raising. Mother animals lead their sad lives between four bars that allow them only to stand or lay down. Turning is impossible. Piglets are warmed by one 100-W-bulb per litter, which is an absurd waste of energy. Investment per mother animal amounts to about 1000 dollars. You can buy more than one hectare of land for that sum in the respective region. Ritter's farm extends over about 2400 ha of land. Like in modern waste management, in agriculture, too, the most unreasonable facilities are promoted.

On the estate of our GAIA-foundation, we are presently experimenting with feeding pigs on wastes from a big slaughterhouse; so far, we have been more successful than expected. Cattle slaughtering yields about 30 kg of stomach contents and 20 l of blood per dead animal. All this has so far been emitted into water, with particles sedimenting in successive lagoons. The first of these lagoons was regularly covered by a crust sufficiently stable to walk on. This crust had to be "disposed", that is, the material – an excellent organic fertilizer – was simply deposited on an illicit landfill.

Today, we cook stomach contents and blood, add a small portion of rice bran, and feed it to the pigs. The other half of the pigs' diet consists of freshly cropped water plants. Now we have founded a little firm, together with the slaughterhouse, especially for the purpose of pig raising. Theoretically, this firm could raise more than 10 000 pigs on the slaughterhouse remainders. We started with a dozen of animals a year ago, at present, their number has already increased to 180. In the days to come, the first of the so raised mother animals will litter.

In the initial phase of our project, we still compost organic remainders until we will have enough pigs. To the extent that the waste load for the sedimentation ponds is relieved, to the extent that they merely receive the rinsing water of the slaughter-house, we will be able to produce an enormous amount of swimming water plants there. This will result in clear, clean water at the outlet of the last lagoon, and additionally in pig food. Meanwhile, we also slaughtered the first of our animals, weighing about 110 kg each. The veterinarian in the abattoir was enthusiasted: no diseases, no parasites. We do not use any antibiotica, medicaments or poisons. Our animals are kept under appropriate conditions, have plenty of space at their avail, are mentally balanced and calm, do not produce liquid manure but only solid one, without flies and stench, which can easily be collected with a rake and be spread on vegetables, fruit or a field after composting. Pigs lie in their own diarrhea only if they are forced to do so, as is the case in modern pig prisons. If kept under appropriate conditions, they never shit near their feeding trough, nor near their sleeping place, nor in their mud bath.

Although this speech focusses on leather, I mention this project because it illustrates a fundamentally different approach. A large conventional waste water treatment facility for the cattle slaughterhouse would cost millions, would create only a few jobs, would serve the capital instead of people, and would destroy nature instead of supporting life and producing clean food.

The water plant sewage facilities are in fact worth mentioning. There are different methods of realizing this idea: basins with swimming plants – particularly adequate for the tropics and subtropics – and facilities operating by means of root area disposal, with upright standing rooted swamp plants, like reed and others, in temperate zones. Organically polluted waste waters can thus be purified in a cheap and efficient way. Associated with this is the emergence of ecotopes of ecological and scenic value – small artificial swamps that attract the entire pond fauna, from coot to stork.

In our state, Rio Grande do Sul, I know a slaughterhouse which would have merited an ecological award. It has established such a swamp. An already very rare species of wild duck is thriving there, the final outlet is absolutely flawless. I came just in time to prevent a catastrophe. The local environmental authority – headed by a chemist – had ordered to drain the swamp. Instead, an expensive conventional waste water facility was to be constructed. This is the way the 'idiots savants' educated in our universities are thinking today! Of course, a chemist does not know anything of biology, let alone limnology.

What is important, of course, is to never let hazardous heavy metals or other, particularly undegradable, organic poisons get into such facilities. As to these, we have to conceive of methods to either renounce their use completely or to hold them back at the source, where they can be reused.

Furthermore, water plants are a means of producing enormous amounts of biomass. In the tropics, up to 600 tons of dry matter per hectare and year, from, e.g., Eichhornia, might be produced on sufficiently polluted waters, in the subtropics about half this amount. This may be used as fodder, may be composted or utilized to produce energy.

Back to leather. For Rio Grande do Sul alone I estimate that leather scraps and leather powder deposited from tanneries onto illicit or more or less organized landfills, together with other urban rubbish, amount to more than 10 000 tons. Remember my introductory remarks. Leather is protein, you see, one of the most valuable substances in building a living organism. In addition, 'wet blue' contains up to 3% of chrome, as you might know, which is presumably more than is contained in chromium ore.

We must not understand these leather scraps as rubbish. They may be untanned. For one year, we have now been cooperating with two tanneries to solve this problem. The recovered chrome can be recycled into the tanning process. The organic constituents of the scraps, liquid aminoacids and protein compounds, may be fed to animals if chrome concentration is below the admissible rate – which we already accomplished in our tests. They can be made into a glue which is applicable in tanneries for certain surface treatments, or they may be used straight in agriculture as organic fertilizer. Besides, we are working in transforming them into a pesticide (no agro-poison).

The remainders of leathers tanned without chrome, like vegetal leather 'wet white' etc. may be used directly in agriculture, for instance as mulch in fruit growing. Their degradation takes quite a while, but once they are degraded, they have the same effect as an organic nitrogenous fertilizer continuously operating over a period of many months. Attention has to be paid to not working vegetal leather deep into the soil, since the contained vegetable tannin inhibits root growth.

Proceeding systematically and systemically step by step, waste and waste water problems of a tannery can easily be solved without much expenses, with cost savings for the firm and biological benefits for society as a whole.

From this point of view, for me as an ecologist tanneries are not necessarily environment-hostile but, on the contrary, to some extent even desirable industries, namely to the extent that they can be integrated into a sound agriculture.

Nevertheless we should consider the possibility of separately depositing leather scraps, since today we are not yet able to recycle all leather rests produced. Separate deposit might allow their later use. I witnessed several times whole truck loads of clean leather rests arriving at the old landfill of Novo Hamburgo, with a barrel of used oil spilled over them or household refuse put on top. Or else leather rests, sawdust, wood ash, plastics scraps, metal cuttings from a lathe, were all thrown together. Here again the mentality "après moi le déluge" if only now I will gain much money! Leather rests are a material that is easily preserved for a practically unlimited period of time without polluting the environment. Several firms could jointly manage a common storehouse for them. In a few years, there will be a market for them.

Later, we will also have to think about how to process used leather objects: shoes, belts, clothes. Due to the diverse dyes and coatings, we are faced with a far more complex problem here. Hence we should even now start reflecting on how to omit hazardous substances in this field. My firm has a small waste sorting facility where 85% of our household and office wastes are accepted. Leather objects, however, are among the 15% which still have to be deposited.

If we want to approach the goals set at the 1992 Environment Conference, we will have to quit single-track thinking. The reason for the non-sustainability of our present economy is that we open ever more cycles that had been closed in nature.

This is particularly true for all further industries producing organic wastes, among them wine cellars, slaughterhouses, sawmills, pulp production plants, food-processing industries, mills and others. Even wood ash is insensibly illicitly deposited in thousands of tons today. Most agronomists seem to have forgotten that wood ash is the best mineral fertilizer you can think of.

I do not mean to focus on industries like furnaces, aluminium smelting works, chemical industry, oil industry, coal industry, metal-working industry, energy and service sector here. Problems in this contexts are sometimes very difficult. If, however, they are approached from an ecological perspective, various aspects might easily be solved which today are said to be insoluble. It should be noticed that a society subsidizing a mad form of agriculture with billions of dollars, which destroys enormous amounts of "surplusses" at the cost of gigantic technical and financial expenses, could well subsidize some sensible recycling projects.

We have to jointly search for solutions. The modern manager has to be aware of the fact that he is not only responsible to shareholders for a sound balance, but also to society, and in particular to Creation as a whole. A sustainable economy can only be an ecological one.

References

Lutzenberger, J. (1994): Knowledge and Wisdom must come back together, Folkuniversitetet, Sweden.

Lutzenberger, J. (1996): We can´t improve nature, in print, Edition Siegfried Pater, Postfach 150106, 53040 Bonn, Tel.: +49-228-236484

Pater, Siegfried (1994): José Lutzenberger, Das Grüne Gewissen Brasiliens (Brazil's green conscience), Lamuv, Göttingen, Lamuv pocket book Nr. 163.

3 Japanese Approaches to Environmental Management

Tomo Shibamiya

History of Pollution in Japan

I should begin by refering to the history of pollution in Japan. Pollution rose in the begining of Meiji era which was the begining of Japanese capitalism 130 years ago. The Meiji government undertook great measures to promote modern industries under the slogan of "inauguration of new industrial enterprises and increase of production".

Pollution in that era was mainly generated by mining. So pollution was basically confined to local areas. After the war, in the 1950s there had been qualitative changes in the Japanese economy. Those changes were

- Energy resources were changed from coal to petroleum. As a result of this change, many petrochemical complexes were constructed. Some complexes were located close to big cities and these have caused SO_x and bad smelling fish. A notorious case of asthma caused by SO_x in Yokkaichi city was called Yokkaichi Asthma.
- As a result of industrialisation, there was a movement of population to the big cities. This movement has led to a rise in pollution in cities through mass consumption and waste water resulting from human activities. Pollution was also caused by vehicle emissions and motor vehicle noise.

Leaders of the industrial enterprises at that time were from petrochemical industries, power companies, steel industries and electronics/electrical industries. Those leaders were well aware of the damages caused by industrial pollution. They started action against pollution.

Let us look at two major instances of pollution, one caused by emissions from petrochemical complexes, the other by discharge of heavy metals in the waste water. They are:

- The Yokkaichi Asthma was caused by SO_x from the Yokkaichi petrochemical complex. The Yokkaichi bad smelling fish was also caused by the waste water from the complex.
- The Minamata Desease was caused by the food chain of methyl mercury in the waste water to a bay from Chisso Company.

The occurrence of this pollution has triggered the establishment of laws and regulations. After the Water Quality Conservation Law and the Industrial Waste

Water Regulation Law were enacted in 1958, the Basic Law for Environmental Pollution Control which is the base of pollution prevention against developing industries was established in 1967. After this, there were many laws and regulations established especially in 1970, when together with revision of above mentioned basic law, 14 pollution related legislations were established.

As a result, laws and regulations concerning pollution control were fully established and the Diet in 1970 was called "Pollution Diet".

In 1971, the Environmental Agency was established for the unification of environmental administration. In the same year, "the Law for the Establishment of a Pollution Prevention Organisation at the Specified Factory with Given Facility" was estabished. This law regulates that a "Specified Factory with Given Facility" must have a designated pollution prevention organisation including a Pollution Control Manager and a qualified engineer for the given facility. Since this law had been established, over 400 000 qualified engineers in various categories were recognised by the government agency and stationed in factories all over Japan.

Local governments have also entered into environmental pollution control agreements containing extreme strict and detailed provisions seeking effective pollution control. Without the permissin of the local government, a company could not construct a factory and start operation in that area. This is a strong measure designed to improve performance and to prevent pollution.

Environmental Management in Companies

As I mentioned above, according to the establishment of those laws and regulations, factories with given facilities are obliged to have a pollution prevention organisation and also measure and monitor emissions and discharges from those factories.

So in the first phase, companies which have some responsibility for above mentioned pollution and leading enterprises equipped with given facilities had started to organise environmental management in the 1960s and through the 1970s. But the environmental management at that time was basically for pollution prevention.

In the 1980s ozone depletion was also recognised by the general public and heavy population concentration in big cities combined with mass consumption have raised a shortage of landfills. Mass consumption has also raised the questions to mass production which means some manufacturer´s responsibility for the mass production of their products.

The recognition of global environmental issues and those pressures on companies led to the re-establishment of environmental management based on the concept of current global environmental protection.

At the same time, many companies which have had no relations with pollution, had also started establishing environmental management systems for global environmental protection.

Countermeasures by Industries

The Japanese industry has been working for pollution prevention together with government administration and has achieved remarkable results. The results can be seen in leading technologies in the world and management systems for resource saving and energy saving.

But, to undertake countermeasures for global warming, the Japanese industry has started to realise that they have to work together for the global environmental protection and sustainable economic development. All of the industry organisations including the Japanese Federation for Economic Organization (Keidanren) have started to work for global environmental protection.

"Keidanren" began by making a team in its environmental committee and starting to work for an environmental charter. As a result of the committee work, Keidanren announced the "Keidanren Global Environmental Charter" in April 1992, as the guideline for industries environmental activities.

Actually, as the announcement of this kind of charter from industry was the first, media reacted greatly to this charter. This charter was also presented in the ICC general assembly in Rotterdam in the same month. In the ICC, there was also an announcement of an ICC Charter.

Major items in this charter are:

- Establishment of environmental management policy
- Nomination of an executive in charge of environmental management
- Establishment of an environmental management organisation
- Documentation of an environmental management system
- Execution of internal audits

This charter is intended to instruct the establishment of an environmental management system in each company. Adding to this charter, the guideline "Ten-Environmental Guidelines for the Japanese Enterprises Operating abroad" was also announced.

One year after this announcement, in 1993, Keidanren made a follow up research how companies were reacting to this charter. They sent questionnaires to 950 major companies in Japan. There was a 57% response rate which is a very high percentage of return for this kind of research.

The answers showed that about 54% of the companies have already nominated executives in charge of environmental management, and about 62% of the companies have already established the organisation for environmental management. This means that 62% of the companies had been executing environmental management in their companies. And also it says that 33% of the companies have established environmental objectives and targets for reduction of environmental impacts, 15% of the companies had been triggered by this charter to establish their objectives and targets.

Regarding the internal audit, 35% of the companies had executed internal audits and additionally 14% of the companies had decided to execute internal audits in future.

So far 33% of the companies have established their environmental management system. The research was done 5 years ago, but the figures may get higher year by year according to recent informations on papers.

Recently, companies and factories adopting environmental management systems based on ISO 14001 are rapidly increasing in Japan. In june 1998, there were 1018 companies and factories certified based on ISO 14001.

The Results of Environmental Management

As the result of the environmental management, Japan has made remarkable recovery from pollution in 1950s and 1960s. I should say there is almost no major pollution from industries. The only major pollution is NO_x pollution by traffic and pollution at lakes and marshes caused by living discharge entering the water. This recovery was achieved by severe observation of laws, improvement efforts by companies and good guidance by government. The results are more than was expected in the past. By improving environmental performance, they could even develop environmental equipment and start environmental business by selling goods such as desulfurisation equipment.

As a result of pressure from shareholders and consumers, and for other reasons, companies are reporting their environmental management in their annual reports. And at the same time, many companies have been issuing pamphlets describing in more detail their environmental management. But very few pamphlets are describing the performance with indices. Recently several companies have started issuing reports regarding their environmental management including environmental performance with indices. Those are the companies who have committed their environmental objectives according to the Voluntary Action Plan to be mentioned later.

This is a new voluntary trend, but, there is a local government which has made a regulation for publication. Firms disclose information which was documented while establishing the ISO 14001 EMS.

I can predict that numbers of companies who will draw up environmental reports and make it public, will increase along with the pressures of interested parties and according to the development of certification systems including ISO 14001.

Also LCAs are now being developed by companies, industrial associations and national organisation. This may not be the result of environmental management, but according to the recognition by the companies. There are 10 major organisations studying LCAs on specific products such as refrigerators, automobiles, basic materials and others.

The Characteristics of Industries Environmental Management in Japan

I do not think that industry is apparently applying Total Quality Management (TQM) as a tool for environmental management, but the basic concept of TQM is being applied. Originally TQM was introduced from the United States as Statistical Quality Control (SQC) and they have developed a Japanese style Total Quality Control (TQC) which includes activities by top management through to all workers. This TQC was spread to all industries in Japan. I think this is called TQM outside Japan.

TQM was called company wide quality control. The following characteristics are listed:

a   Thorough quality first in business management
b   Policy development and management of the policy
c   Quality assurance activities in the planning and development division through to the sales and service division
d   Spreading of TQC from manufacturing sector to various sectors
e   All employees' participation
f   QC training
g   QC circle activities
h   QC diagnosis
i   Making use of statistical methods
j   Nation-wide QC promotion activities

As Japanese top management has recognised the importance of pollution prevention by learning from the penalty of pollution, they have done their best to prevent pollution. At the same time, there was a change in the recognition toward industries by the public, that means that a company which has caused pollution will be subject to serious damage of its company image.

On the other hand, to pursue businesses, companies have been introducing policy development and management by "management by objectives". This kind of management system is especially strong in the quality assurance area.

As I mentioned earlier, companies have changed the scope of their EMS from pollution control to global environmental protection. Soon after, the Earth Summit in 1992 again triggered companies to establish their long term policies and objectives such as energy reduction as countermeasure for global warming. The database for those objectives has been made from the results of their past environmental management.

Along with the international committment by government to reduce CO_2 emissions, there was a request to companies from government regarding environmental improvement. It was called "Voluntary Action Plan for Environment". The major request from the government was to establish environmental improvement objectives in indices. As a result of the request, many major companies have established their objectives and made them public. Six companies including Toshiba and Hitachi in the electrical and electronics sector have developed standard units for each performance objective such as energy consumption per unit of sales (kWh/ million Yen).

And in the spring of 1992, executives in charge of the environment in those six companies made a press release as a jump start for industries which gave rise to a significant level of response.

By the Voluntary Action Plan., a company is asked to establish a set of long range environmental objectives such as to stop using CFCs, energy saving, reduction of solid waste or products improvement. To pursue those objectives, the company has to nominate an executive in charge of the environment, to establish and improve its environmental organisation, pollution prevention system, awareness and training program and other means. These committments were established by sincere top management recognition of the environment.

By beginning, of 1996 about 350 major companies had established and made public their "Voluntary Action Plans" according to the investigation of Keidanren. Some of the companies have started publication of the performance results. I think this number is quite high because those 350 companies have also some kind of management systems in their organizations.

Employee Participation in TQC

Quality Circle activity has been adopted since the 1960s and since then this activity has been carried out in almost all major companies in Japan. Quality Circles are now organised in every department of the company, not only at the production line but also in the engineering department including the environmental related employees.

There are quality circles which are involved in discussing environmental issues related to the groups and successful improvement results have been reported from the groups.

Training on quality control has beeing done for many years, and there is also training for how to use quality control tools such as fish-bone chart, X-R charts and others. These statistical tools are also being used in the quality circle activities even when solving the environmental issues. In the case of waste water treatment facility operation, operators use such tools for establishing and improving self-imposed control standards for the facility.

There are a couple of good examples for employee participation. Companies have started to use environmental notice boards or "Greenboards" to make their employees aware of their environmental policy, objectives and activities in front of the company entrance. Other Greennboards are situated in break rooms and designated areas in each department.

There are other examples of intensive employee information in the area of solid waste storage and environmental facilities. Visual displays are used instead of documented instructions for all employees to understand and participate in environmental protection activities.

As I explained before, it is not apparent that TQM is applied in the environmental management, but in the basic way of operation they utilise the TQM methodology. I have shown fragmentary examples of environmental management in Japan. Because of the similarity of the principle, an environmental version of TQM will be formed soon.

Quality certification has begun quite recently, and the number of certified organizations is 2000 by the latest statistics. I think this number is far behind that in European countries.

In the environmental area, the Japanese Accreditation Body, which is abbreviated JAB, started accreditation in June 1996. It also started auditor qualification systems at the same time. This accreditation body also handles quality. Until june 1998, there were about 1018 organizations with certified environmental management systems by certification bodies.

From our short certification experience and other industry information, we in JACO think, that so far the standards themselves are different, and people in charge of Quality and Environment are also different. For some time, the certification will be done separately for both the auditor and auditee side.

How Do International Discussions on EMS Standards and EMAS Have an Effect on Japanese Companies?

As I mentioned before, the management of Japanese major companies is well aware of the necessity of environmental management systems for pollution prevention and global environmental protection.

When ISO TC 207 started in Toronto in 1992, Japan sent its delegation with about 40 members. Keidanren sent 20 members from industry out of 40. Since then Keidanren has been sending its representatives as committee members, subcommittee members and working group specialists. Among these industry members, 1/3 of them are from the electrical and electronics industry.

This means that management of electrical and electronics industry are well aware of the needs of EMS from its experience in the past, the nature of its export businesses and wider descernment through their businesses. They also have much more concern regarding EMAS, because it may have some effects on companies operating inside as well as outside of EU.

Recently 10 major electrical and electronics companies established a third party certification organization which is named Japan Audit and Certification Organisation for the Environment (JACO). JACO has started its business last in 1995 and it has already certified many organisations. Since ISO 14001 came into effect, many companies and factories are applying to JACO. According to papers, many major companies are announcing that they are going to get certificates until 1999. This is especially the case for the electrical and electronics industry which has been exporting business to Europe and has factory operations in EU and look to get certificates earlier.

Conclusion

To conclude I should say that the purpose of the environmental management in Japan is to pursue performance improvement and to achieve certain objectives. Industry is using the TQC way of doing things which ranges from top management committment through to all employee participation.

As I mentioned, my company JACO is going to support all industry who want to have environmentally conscious operations, according to its competence. JACO is going to have operations not only in Japan ,but also in other countries including South East Asia.

4 Reducing Environmental Harm from Products: More than Selling "Green Goods"

Frans Oosterhuis

Introduction

Economists always start their analysis with "Let's assume...". The writer of this chapter is an economist, so let's assume we are a producer and we want to reduce the environmental damage caused by the product we make.

In fact, this is an assumption which a "traditional" economist is not very likely to make. In his or her view, a producer is a profit maximising, self-interested individual who will never contribute voluntarily to environmental improvements. Why? Because the costs of doing so would have to be borne entirely by the producer, whereas the environmental benefits would accrue to society as a whole.

Such "irrational" behaviour is anathema to an orthodox neo-classical economist. Instead, he/she would say: "Let's assume we are a policy maker who wants producers and consumers to reduce the environmental damage caused by the products they make and buy". Indeed, most of the economic literature dealing with environmental policy takes that point of departure. It takes the perspective of the impartial, benevolent authorities who should interfere in the market so as to reconcile "rational" producer and consumer behaviour with optimal environmental improvement.

In reality, we observe a different picture. Policy makers and enterprises, as well as other actors (interest groups, consumers etc.) all have their own set of objectives and demands concerning products. Increasingly, environmental aspects are part of those sets.

They have various sets of instruments with which they can try to pursue those objectives. In the policy maker's tool box, we will find traditional regulatory instruments, such as product standards and bans, as well as economic instruments, such as product taxes. Producers may use their own means to promote their product-related environmental objectives, e.g. giving their products a "green" image in their marketing campaigns, or providing environmental information on the label. Many instruments can be applied both by the government and by enterprises. Information provision, take-back schemes and standards can be either prescribed by regulation, or applied voluntarily by producers at their own initiative. Moreover, governments and firms often engage in voluntary agreements to achieve certain environmental targets. Even environmental organisations nowadays promote co-operation with enterprises in their pursuit of environmentally less harmful products.

In this chapter, we will have a look at the new role of firms as suppliers of environmental improvement through "greener" products. It will become clear that the production of less polluting products is only part of the story. Enterprises also have to pay more attention to those stages of the product's life cycle which are traditionally beyond their control. Moreover, they should consider the functions which their products fulfil if they really want to reduce the overall environmental impact. We will conclude that policies are needed which stimulate firms to accept extended responsibility for their products and to reduce the "product intensity" of the economy.

Products and the Environment: Strategies and Policies

In the developed market economies of Western Europe, comprehensive environmental policies emerged around 1970. At that time, such policies were characterised by an orientation towards single media (air, water, soil) and a top-down approach, with detailed, government-initiated regulations (command and control). These aimed to obviate the most urgent, visible and obvious problems. These early environmental policies can be called *process policy* (focusing on the production stage, "end-of-pipe" oriented) and *waste policy* (focusing on the waste stage, "end-of-life" oriented).

Despite some success in reducing emissions, the limitations of these policies became obvious. They were not suited to addressing complex environmental problems in an integrated way, which would take into account all substance flows involved as well as their intricate relationships. Moreover, adequate enforcement appeared difficult to implement, and "non-point" sources of pollution, such as products, were more or less neglected. Gradually, more comprehensive approaches were developed. One of these is *product policy*, which, ideally, addresses all environmental impacts of a product during its entire life ("cradle to grave").

Environmental authorities in the European Union and its Member States are just starting to discover the possibilities and limitations of such an integrated approach. The most obvious policy instrument based on a multidimensional environmental assessment of products is the eco-label. Its origin lies in Germany in the late 1970s, and in recent years it has been applied in a large number of countries, as well as at the EU level. The application of other policy instruments which would encourage the supply of relatively "clean" products and discourage the relatively "dirty" ones is still rare. Examples would include lower tax rates for the "cleaner" goods (e.g. VAT differentiation), and the use of environmental criteria in decision-making on public expenditures. If such instruments are used at all, they usually focus on one environmental aspect (e.g., reduced excise rates for unleaded petrol), rather than on all aspects over the full life cycle of the product.

In the meantime, awareness among enterprises has been growing that the environment is not just a set of regulations imposed by bureaucrats, which should be opposed or at best reluctantly obeyed at minimum cost. Firms recognize nowadays that the demand for clean air, water and soil is as real as the demand for the products they make. Indeed, consumers as well as authorities and other actors want them to supply products which not only provide satisfaction in terms of their primary functions for the user, but also in terms of their impacts on the environment during their entire life cycle.

Companies have responded to these new challenges in different ways. Some of them have persisted in a reactive strategy, changing their behaviour only when it was unavoidable. Many others, however, developed new "pro-active" strategies, in which the environmental challenges were to be transformed into opportunities for product innovations, the development of new markets and a re-thinking of established production and distribution processes. New concepts came into being, such as "life cycle assessment" (LCA), "design for environment" and "eco-design".

The development and marketing of "greener" products is nowadays already second nature to many firms. Innovating products has become a necessity of life for most industries, and environmental requirements can simply be regarded as an additional item on the list of criteria which the new product has to fulfil. Alleged environmental superiority can even be used to distinguish a new product in a market for otherwise rather homogeneous goods.

So far, so good. But selling a "greener" product is not exactly the same as saving the environment. A large part of the life cycle of the product is usually not controlled by the producer. He/she can influence the production stage and in many cases also the preceding stages (by imposing environmental requirements on suppliers), but once the product has been sold, its environmental impact depends on the buyer's behaviour. Careless use, inadequate maintenance and improper disposal can turn even a "green" product into an environmental threat. The fact that a product is "environmentally-friendly" may even lead people to treat it more negligently. A well-known example is the cleaning of brushes which have been used for water-based paint under the tap, whereas they should be treated as chemical waste, just like solvent-based paint. Likewise, energy-efficient goods may induce people to use them more frequently or longer. This is known as the "rebound effect". Moreover, the increase in consumption can easily offset any reductions in the amount of pollution per product. Therefore, a product-oriented environmental policy cannot restrict itself to stimulating the production and consumption of environmentally preferable goods. It also has to address issues with which the traditional enterprise feels less familiar: how to make sure that people will use, maintain and dispose of products in an environmentally sound way, and how to control the increase in the amount of products consumed.

Obviously, environmental authorities and consumers have their own responsibility with respect to these issues. But here we will look at the role of the enterprise. How can they contribute to an environment-oriented "after sales management"? Should they accept responsibilities for the environmental fate of products which they made but do not own any more? Should they make consumers buy less? What are the risks and rewards of such an approach? And how can policy makers assist and stimulate them?

Box 1. Exemplary products: paint and batteries[a]

In the EU, some 2.5 million tonnes of decorative *paint* are produced and consumed annually. About one third of this is based on organic solvents, resulting in VOC emissions of 350 000 tonnes per year. These emissions are the main focus of paint-oriented environmental policy and product development concentrates on low-solvent alternatives. However, other stages in the life-cycle of paint also have environmental relevance, e.g. raw materials (such as TiO_2) and waste (such as residual paint and old paint layers). Reductions in the environmental impact of paint use can be achieved, among other ways, by reducing the frequency of painting and by skilful application. This could be stimulated, for instance, by including the durability of paint in the criteria for an ecolabel, and by introducing an eco-label for the services of professional painters. Other promising instruments are a levy on the paint's solvent content and (voluntary or obligatory) take-back of paint cans by the supplier.

More than 3 billion *batteries* are sold in the EU each year, of which more than 90% are non-rechargeable. The main environmental concern with batteries is their heavy metals content (in particular mercury, lead, cadmium, zinc and nickel). Policy attention is focused on reducing the heavy metal content and on the waste stage of the product: measures are taken, with varying degrees of success, to ensure the collection of spent batteries and their environmentally safe disposal or recycling. Reducing the amount of batteries used is hardly stimulated, except by promoting the use of rechargeable (e.g. NiCd) batteries. However, this substitution can only be beneficial in environmental terms when the recovery of the used cadmium is ensured. Meanwhile, recollection rates of spent batteries are disappointingly low. Switzerland, where battery users are legally obliged to bring back batteries containing hazardous substances, has the highest return rate in Europe: 60%. Financial incentives, such as return premiums or deposits, are probably needed to achieve higher rates.

[a] This text is based on Oosterhuis et al. (1996), Sections 4.2 and 4.3.

Extended Producer Responsibility

If producers really feel responsible for the environmental impact of their products, there is no reason why this responsibility should stop at the delivery outlet of the factory. Many firms already accept to some extent an extended product responsibility or "product stewardship". This acceptance is often stimulated by (the prospect of) regulation. A growing number of initiatives to take back products after their useful life can be observed, e.g. in the area of batteries and electronic appliances. Many companies also provide (voluntary or compulsory) information on the product label which should enable the consumer to use and discard it in an environmentally compatible manner.

Product design which facilitates re-use and recycling can also be regarded as an expression of accepted extended producer responsibility. Examples are the marking of plastic parts with a view to recycling, and the simplification of disassembling complex products, so as to enable the user to replace parts rather than the whole product. Some firms (e.g. copier manufacturers) even "remanufacture" used parts, resulting in waste reduction as well as financial savings.

Despite these various actions, there are still a lot of opportunities for improved "after sales management" with regard to minimising the environmental harm caused by products. A crucial role in this respect is played by the retailers, the intermediate link between producer and consumer. Retailers can provide the consumer with information and instructions regarding environmentally safe product use and disposal. This is especially important for information which is less easily conveyed by written means. Moreover, they can communicate experiences, questions and complaints from the consumers to the producers.

Furthermore, producers can provide incentives for environmentally sound product use and disposal. The most well-known example of such an incentive is the deposit-refund scheme. These schemes have a long standing tradition in the area of packaging, in particular glass bottles. Environmental considerations have led to a revival in the interest for deposit-refund systems. In several European countries, they are now also applied to other drink packaging, such as PET bottles and aluminium cans, as well as to scrap cars. Other areas suitable for such application have been suggested, including batteries. Deposit-refund systems can be imposed by the environmental authorities, but the preferred option is to initiate them on the basis of co-operation with industry. A problem with these schemes is that when they reach a high return rate, the difference between deposits received and return premiums paid is often not large enough to finance the administrative and handling costs.

For durable goods, overhauls may be good occasions to maintain or even improve their environmental performance. In the case of cars, emissions already have to be checked regularly in many countries. Such checks could be extended to other durable goods and include, in addition to emissions, other environmental features, such as energy consumption. Regular and proper maintenance may also lead to a longer product life, which is usually environmentally beneficial in itself.

New environmental insights can lead to bans or restrictions on the use of certain substances and materials. Usually, these regulations apply to new products (or new types) only. The existing stock of products containing the dangerous substances can remain in use for many years or even decades. Examples are transformers and capacitors containing PCBs, and refrigerators and fire extinguishers containing halogenated hydrocarbons. To prevent uncontrolled release of the substances into the environment, early replacement (either of the whole product, or, when feasible, of the substance only) could be considered.

From Product Fetishism to Service Orientation

Consumers consume because they want to avoid hunger, pain, cold, danger and suffering; they want to enjoy beauty, pleasure, health and convenience; they want entertainment and spiritual and physical development; they want communication and companionship; and they want all these things for their children as well.

Producers produce products and services which should help to meet these wants and needs. However, in a market economy the producer has to focus on the profitability and continuity of the firm. In order to earn sufficient return on investment, the capital stock should be used at the fullest possible capacity. It seems that producers who stimulate their customers to reduce consumption in the interest of the environment dig their own graves.

In reality, this conclusion is too straightforward. Durability and longevity can be strong marketing arguments, and the supplier of hard-wearing products can sometimes gain a market share large enough to offset the lower frequency of replacement which reduces turnover. Nevertheless, in most industries there is a "natural" tendency towards the maximisation of production volume, given the installed capacity.

One of the few areas where suppliers have started to advise their customers on reducing consumption is the energy industry. Since the first oil crisis in 1973, energy conservation has been promoted not only by governments but also by the suppliers of energy, in particular electricity distribution companies. This exceptional behaviour can be explained by the fact that these companies are often government-owned or at least under intensive public control. Moreover, until recently they had a legal monopoly in the largest part of their market, so that they could recover additional costs and lower revenues resulting from their energy conservation activities. Meanwhile, this situation is changing due to the liberalisation of electricity markets in the EU.

Will it be possible to copy this model from energy to other economic sectors? Are the sales(wo)men of home appliances going to recommend us not to buy dishwashers and dryers but to wash the dishes manually and hang out the washing in the sun, with a view to saving energy and natural resources? It seems highly unlikely. However, the example given by energy companies can at least teach some lessons.

First of all, enterprises have to realise that in many cases it is not the product itself which the customer is interested in, but the function that this product fulfils. The next step is to explore possible alternatives for fulfilling that function, which are less polluting and/or require less energy and materials. In other words, the emphasis has to shift from the product to the service. Sometimes, the sale of a new product will be part of that service; in other cases, the consumer could be provided with alternative means of meeting his/her wishes, e.g. renting or leasing the product; repairing or upgrading an existing product; or taking measures to remove the need for the product.

Box 2. Leasing, sharing and pooling[a]

Improved resource efficiency causes a reduction of substance flows, which can be achieved *through* two approaches (Stahel 1994, p. 201):

- *a prolonging* of the utilisation period of products and
- *an intensifying* of the utilisation of products.

   Both approaches can be subsumed under the heading of *dematerialisation*. Some typical innovative instruments for dematerialisation are:

Eco-leasing: products are no longer purchased by the customers but given for usage over a certain period (Hinterberger et al. 1994, p. 19). During this period the customers have to pay a rent, and the producer earns more the longer and more intensively the product is used (Zundel et al. 1993, p. 23). Through this kind of rearrangement of ownership it is in the interest of the producer to supply durable and repair-friendly goods that are easily recyclable after final use. Eco-leasing is mainly relevant for durable goods (cars, computers, copiers etc.), but there are also examples of leasing schemes for non-durable goods, such as solvents (Hockerts et al. 1993, p. 20f).

Sharing and pooling: the general idea behind these instruments is that products which are not frequently used need not be owned by the user. The shared and intensified use of such products could reduce overall product throughput and, hence, reduce waste. The *simultaneous* shared use of durable products is already common practice, e.g. in the area of public transport. Interest in new forms of *successive* use of the same product by several owner-users is growing. Car sharing schemes are a prominent example. Apart from the environmental benefits, the advantages of such scheme are lower fixed costs and the opportunity for users to choose the most suitable alternative out of a given pool (cf. Hockerts et al. 1994, p. 10f).

[a] This text is based on Oosterhuis et al. (1996), Section 7.4.

   If a company decides to take this service-oriented approach, it may be a good idea to separate the business unit which makes the product from the one which provides the service. This allows both units to operate on a commercial basis and gives the "service unit" the opportunity to look for the best solution for the customer, without the need to make the sale of the company's product part of that solution. This way of operating is much the same as is already common practice in many vertically integrated enterprises.

New Approaches for Product Policy

At the beginning of this chapter, we started with the assumption that we are a producer and we want to reduce the environmental damage caused by the product we make. We have seen that, in addition to making and marketing "green" products, more attention should be paid to the "after sales" stages. Our responsibility as a producer should be extended to include the environmentally sound use and disposal of our products, as well as a reduction in the amount of the product sold and used when the functions they fulfil can be performed by less material-intensive means.

The question then remains, as to how this can be achieved in a competitive market economy. To some extent, the shift from a product oriented to a service oriented approach may prove a profitable one, even under present market and regulatory conditions. This will particularly occur in cases where consumers have an information deficit which lead them to spend too much money on a product instead of choosing less costly and environmentally preferable alternatives. The existence of many unutilised opportunities for profitable energy conservation shows that net benefits can be achieved by providing information on how to reduce the consumption of a product (in this case energy) while maintaining or even improving welfare. Such situations might well exist with respect to other products as well.

But apart from these "win-win" options, our enterprise will have to face difficult choices in cases where the immediate profits from selling a new product are likely to be higher than the revenues from maintenance and repair or from assistance in the provision of alternative ways of fulfilling the function demanded. As long as the costs of pollution and waste which a product causes are not fully included in its price, the market keeps sending biased signals. The result is a continuation of the present dominance of product sales at the expense of service and function orientation.

Collective solutions, which would enable us to gain from a more "dematerialised" approach therefore deserve our support. Such solutions could take many forms, ranging from classical government regulation, through economic instruments such as charges and deposit-refund systems, to voluntary agreements and self-commitments. Generally, the measure which gives the strongest incentive while leaving maximum room for choices will be preferred. More important than the instrument chosen, however, is the recognition that a firm which takes its environmental mission seriously should be in favour of policy reforms which make it commercially attractive to shift towards environmentally benign product treatment and reduced product "throughput".

References

Hinterberger, F. et al. (1994): Increasing Resource Productivity through Eco-efficient Services. Wuppertal: Wuppertal Paper no. 13.

Hockerts, K.; Geissler, F.; Seuring, S.; Petmecky, A. (1993): Kreislaufwirtschaft statt Abfallwirtschaft. Bayreuth: own publication (Interim report).

Hockerts, K.; Seuring, S.; Petmecky, A.; Hauch, S. and Schweitzer, R. (eds., 1994): Kreislaufwirtschaft statt Abfallwirtschaft. Ulm: Universitätsverlag.

Oosterhuis, F.; Rubik, F.; Scholl, G. (1996): Product Policy in Europe. New Environmental Perspectives. Dordrecht/Boston/London: Kluwer Academic Publishers.

Stahel, W. (1994): Langlebigkeit und Mehrfachnutzung – Wege zu einer höheren Ressourceneffizienz. In: S. Hellenbrandt and F. Rubik (eds.), Produkt und Umwelt. Anforderungen, Instrumente und Ziele einer ökologischen Produktpolitik. Marburg: Metropolis.

Zundel, S.; Arndt, H.-K.; Leinkauf, S. and Sartorius, C. (1993): Elemente volkswirtschaftlichen und innerbetrieblichen Stoffstrommanagements (Ökoleasing, Chemiedienstleistung). Berlin: Institut für ökologische Wirtschaftsforschung.

5 Environmental Benefits Through Company Application of LCA

Gerd Ulrich Scholl, Susanne Nisius

Introduction

Life cycle assessment (LCA), i.e. the systematic inventory and evaluation of environmental impacts of a product "from the cradle to the grave", is an emerging tool. On the one hand it is used by governments, e.g. when establishing ecolabelling criteria for certain product groups or when defining mandatory re-use or recycling quotas, as it has been done in the context of the German packaging ordinance. On the other hand they are increasingly applied by companies for the identification of environmental weak spots in products and for product development. More information on environmental product policy in Europe is given in Oosterhuis/Rubik/Scholl (1996) and Grotz/Scholl (1996a). A comprehensive and very detailed overview of the different applications of LCA is given in Curran (1996) but has the use of this new and promising tool ever yielded environmental improvement?

The German Ecological Economics Research Institute (IÖW) has addressed this question in a project funded by the Ministry of Environment and Transport of Baden-Württemberg. The core of our research has been the description and evaluation of eight case studies from different industrial sectors (Table 1) of which two will be briefly presented below.

The Food Example – "Neumarkter Lammsbräu"

The "Neumarkter Lammsbräu" is a small brewery that principally supplies regional markets in Bavaria. It has about 80 employees and an annual beverage output of 8 000 000 litres. Its main product is beer from raw materials that are organically grown. It is the market leader among German suppliers of eco-beer with a market share of 60%.

Due to the initiative of its managing director, the environmental profile of activities was first documented in a comprehensive ecobalance in 1991. According to an IÖW materials accounting methodology this is divided into a "factory-", "process-", "product chain-" and "location-balance". On the basis of this first ecobalance

Table 1.   Characterisation of case studies

Company	Industry Branch	Subject of the LCA
AEG Hausgeräte GmbH, Nürnberg	Electronics	Vacuum cleaner tube
Augsburger Kammgarn Spinnerei AG, Augsburg	Textile industry	Yarn
Junkers, Subsidiary of Bosch GmbH, Wernau a.N.	Electronics	Transport packaging
Byk Gulden AG, Singen	Chemical industry	Packaging
Donau Tufting GmbH & Co KG, Denkendorf	Textile industry	Carpets
Grammer AG, Amberg	Furniture	Office chair
Henkel KGaA, Düsseldorf	Chemicals and Consumer goods	Laundry detergents (tensides)
Neumarkter Lammsbräu, Neumarkt	Food	Beer

an eco-controlling system has been installed which is aimed at implementing the achievement of agreed environmental targets.

The following exemplary weak spots were identified in the framework of the first ecobalance along the product chain (see Table 2): insufficient fraction of raw materials (barley and hop) from organic cultivation, sub-optimal allocation of transport, use of chlorine and phosphorous containing cleaning agents and stoppers containing PVC. The "Neumarkter Lammsbräu" afterwards introduced certain improvements and a few examples follow.

The brewery succeeded in completely shifting to organic inputs by changing their suppliers and/or their suppliers way of cultivation. Thereby, the CO_2 emissions caused by the production of chemical fertilisers may have been reduced by approximately 60%. Moreover, the dissemination of pesticides into the ecosphere may have been avoided and traces of cadmium and nitrate in the end product minimised. Unfortunately, these effects cannot be further quantified. Other optimisations refer to an improvement of the cooling cycle and the increased use of solar energy. The firm has not been able to change, for instance, to an environmentally more sound fleet of transport vehicles and to install a "greener" CHP plant.

This ecobalance in 1991 was the first step by "Neumarkter Lammsbräu" in a process of continued improvement. This process covers organisational issues (e.g. establishment of an environmental department and appointment of decentralised eco-representatives), technical aspects (e.g. computer-based documentation of environmental weak spots) and questions of information (e.g. a system of environ-

Table 2. Summary of the case study "Neumarkter Lammsbräu"

Results of the LCA	Measures for Optimisation	Potential Environmental Benefits	Actual Environmental Benefits
Insufficient fraction of barley and wheat from organic cultivation	Shift to inputs from organic cultivation	Reduced air emissions (CO_2)	Reduction of the specific CO_2 emissions by 60%
		Renunciation of pesticides	Reduced immission of pesticides
		Lower nitrate and cadmium values in the final product	Lower nitrate and cadmium values in the final product
Environmental burdens through oil heating	Substitution of the oil heating through a "greener" CHP plant	Reduction of air emissions (CO_2)	None, because not (yet) realised
		Reduced consumption of non-renewable resources (mineral oil)	
Insufficient use of waste and rain water	Increased use of waste and rain water	Diminished water consumption	Partial, because not completely realised
Insufficient use of environmentally friendly transport media	Improvement of the fleet (e.g. use of catalytic converters for diesel cars)	Reduced air emissions	None, because not (yet) realised
		Reduced consumption of mineral oil	
Use of toxic cooling agents (e.g. ammonia)	Optimisation of the cooling cycle (through cascades)	Reduction of environmental risks (e.g. CFC, ammonia)	Partial, because not completely realised
		Reduced energy consumption	
Environmentally harmful energy supply	Use of solar energy	Reduction of non-renewable energy use	Yes, because completely realised

mental performance indicators). The environmental benefits generated by these measures can only indirectly be assigned to the product chain balance or to product-related aspects of the overall ecobalance. They will be illustrated by selected environmental performance indicators. The specific water consumption and specific amount of waste water, for example, was reduced steadily during the last couple of years. The water consumption was curtailed from 12 hl per litre of beer sold in 1989 to 7 hl in 1994, i.e. by more than 40%. The amount of waste water was reduced accordingly by approximately 50%.

The Packaging Example – "Bosch-Junkers"

The main business of "Bosch-Junkers" is production and trade in heating- and water heater equipment. It is a subsidiary of the "Robert Bosch GmbH" since 1932 with production plants in different European countries. In 1995 the company had approximately 1800 employees.

In 1992 the company commissioned a consultant to carry out a comparative LCA of one-way transport packaging (OWTP) and returnable packaging (RP) for heating devices. The OWTP consists of cardboard with a wooden frame and a polystyrene cushion. The RP is a folding box made from polypropylene and polyethylene. The objective of the study was to compare the advantages and disadvantages of both alternatives and to improve the environmental performance of the RP.

The LCA comprised the inventory stage followed by a verbal-descriptive interpretation. The functional unit was the distribution of 1000 heating devices. Since there were no experiences with RP, scenarios of 5, 10, and 20 return cycles were assumed. The results and the ensuing measures taken are summarised in Table 3. In the following we will highlight a few examples.

The LCA identified the colouring agents contained in heavy metals present in RP as a weak spot. As a consequence, the next series of RPs was produced in environmentally more benign grey colour. By the avoidance of heavy metals environmental benefits could be achieved during the final disposal of packaging. However, due to a lack of data we were not able to quantify this impact.

A reduction of the one-way parts of the RP could be realised as well. Although the polyethylene (PE) bag is still necessary, one could avoid the use of cardboard interfacing and a polypropylene (PP) band. The resources saved, (cardboard, mineral oil), could not be calculated due to a lack of data.

Furthermore, the LCA found that the use of recycled plastic for RP could reduce the consumption of raw materials, pre-manufactured goods and energy. One could save, for instance, 44% of crude oil, 44% of hydrogen, 37% of auxiliary materials, 29% of water, and 44% of heating gas. Moreover, a curtailing of atmospheric emissions and effluents was possible. Hence, there was great potential for environmental improvement, but it has not been realised so far. The main reason was that the quality of plastic needed was not available in sufficient amounts.

There have been benefits reported by substituting OWTP with RP at the level of life cycle inventory depending on the kind of material observed. Net benefits that are independent of the number of return cycles occur with cardboard, wood, and polystyrene. In a pilot project started in the region of Hanover they saved 20 tons of cardboard by avoiding 4835 OWTPs within the time period from 1993 to 1995. In other material categories, such as plastics, there were additional burdens arising through the manufacture of RP. These negative impacts, however, correlate with the number of return cycles, i.e. the more often the packaging is re-used the smaller is the specific burden. In this respect the LCA finds that RP is environmentally superior to OWTP when it is returned more than eight times. For the time being,

Table 3.   Summary of the case-study "Bosch-Junkers"

Results of the LCA	Measures for Optimisation	Potential Environmental Benefits	Actual Environmental Benefits
RP environmentally superior (up from a return rate of eight)	Consultations on a branch level	Reduced waste

Reduced use of raw materials (e.g. paper, wood) | Saving of 20 tons of paper in the pilot project

Increased use of plastics in the pilot project |
Use of heavy-metals containing yellow pigments	Manufacture of RP in grey colour	Reduced emission of hazardous substances during after-use management	Realised, not quantifiable
Avoidable one-way parts of RP	Reduction of one-way parts	Saving of resources, reduction of wastes	Realised, not quantifiable
High energy consumption for the production of PE/PP	Use of recycled plastics	Reduction of energy consumption	Not realised
Environmental burden through long transport distances	Solution on a branch level	Reduction of transport-related energy consumption and emissions	Not successful

one should note that, since RP does not reach such return rates in the pilot project, its substitution for the OWTP still causes additional net burdens. Environmental benefits could therefore be achieved, if the pilot project were carried out on a larger scale with higher return rates for each RP.

Our analysis of the "Bosch-Junkers" LCA has shown how difficult it is to realise possible environmental benefits. The potentials in this example are obviously rather large; the actual improvements, however, quite small. It appears that the translation of results and recommendations of an LCA into practice works well, if this is feasible by the company itself (e.g. alteration of the colouring of the RP). The dissemination of RP in the market was however not successful due to a great number of different actors necessarily involved (e.g. competitors, retailers, plumbers) and their diverging interests.

LCAs and Patterns of Application

The study has revealed that the application of LCA in industry can imply environmental benefits. Whether it actually does, is strongly dependent on the objective of the LCA and its pattern of application. We distinguished two typical patterns.

On the one hand, LCA is applied in a "retrospective" pattern that illustrates the success of environmentally sound product development in retrospect by demonstrating, for instance, that optimisation of production processes has led to a curtailing of the energy content of the product under consideration.

On the other hand, a "prospective" application of LCA serves to identify environmental weak spots and ensuing optimisation of the product. This is the case when the assessment of an office chair, for example, has revealed the use of environmentally harmful substances and when these substances are avoided in follow-up series of the chair. It is obvious that the latter way of application generates more potential environmental benefits than the first.

In our eight case studies of the practical application of LCA in companies we rarely observed the second "prospective" pattern. Accordingly, the actual environmental benefits occuring in our sample have been quite modest. The result is further due to the fact that most examples we analysed have been pilot projects of LCA application. In these cases environmental benefits might arise through follow-up projects when the enterprise has become more familiar with this new tool and when its application is more deeply embedded in the company context.

Supporting and Hindering Factors

The case studies have shown that a plethora of different factors influence results and realisation of LCA recommendations. They can be assigned to the categories "costs and benefits", "information and communication", "organisation" and "technology". A few examples of the relevant determinants we have identified will be presented in the following.

If there are potential environmental benefits, economic aspects take an essential role during the realisation of actual impacts. Expected savings or future returns can exert a positive influence. Possible financial benefits, however, are often indirect e.g. positioning of the company as an "eco-pioneer" in the market in order to ensure its long term economic existence. On the other hand, economic aspects appear to play a minor role during the implementation of a LCA, i.e. "learning costs" for becoming experienced with the methodology are usually no impediment.

An active communication policy about and by means of an LCA apparently supports the realisation of environmental benefits. When conducting a LCA the intensive exchange of information, internally within the company for data collection and processing, and externally between the enterprise and consultants, suppliers, traders etc., is of great importance. A more intense co-operation can contribute to improving the feasibility and acceptance of ensuing recommendations and to establishing a kind of "life cycle thinking" within the firm.

Moreover, the achievement of environmental improvements is dependent on the way the LCA is embedded into the company's organisational profile. In-house interdisciplinary work panels, for instance, working closely together with the consultancy can create a feedback back to their specific departments and, thereby, positively influence the following innovation process.

Furthermore, we observed that in the course of increasing LCA application companies go through a learning process that starts with pilot projects, followed by the establishment of internal knowledge and eventually leading to a more systematic and prospective way of application, e.g. in the context of green product development. This process may generate more environmental benefits in the long run.

Conclusions and Outlook

Our working hypothesis that the application of LCA in industry leads to environmental benefits has been verified carefully. In order to achieve eco-benefits many obstacles have to be identified and overcome.

An important result of our research has been that the theoretical ideal of an LCA, such as called for in the international standard ISO 14040, is rarely encountered in practice. Therefore, the formulation of minimum standards for a "streamlined" LCA that help overcome the dilemma between scientific precision and practical applicability is needed (Cp. Weitz et al. 1996). Furthermore, the application of LCA can be fostered on a larger scale by increased dissemination of suitable software tools and improved access to databases that provide information, e.g. on the primary energy consumption of selected materials.

Moreover, LCAs should be more deeply embedded in decision-making processes of the company. This could be achieved, for instance, through a "dynamisation", i.e. the frequent conduction of abridged LCAs in order to establish product-related environmental performance indicators, or through further integration into environmental management schemes.

References

Curran, Mary Ann (1996): Environmental Life-Cycle Assessment, New York et al.: MacGraw-Hill.
Grotz, Susanne; Scholl, Gerd (1996a): Application of LCA in German Industry, in: The International Journal of Life Cycle Assessment, Vol.1 No.4 1996, pp. 226–230.
Oosterhuis, Frans; Rubik, Frieder; Scholl, Gerd (1996): Product Policy in Europe: New Environmental Perspectives. Dordrecht, Boston, London: Kluwer Academic Publishers.
Weitz, Keith, A.; Todd, Joel, A.; Curran, Mary A.; Malkin, Melissa, J. (1996): "Streamlining Life Cycle Assessment. Considerations and a Report on the State of Practice.", in: International Journal of LCA, Vol.1, No.2, pp. 79–85.

6 Competitive Advantages Through Voluntary Environmental Reporting

Klaus Fichter, Thomas Loew and Jens Clausen

Growing Importance of Environmental Reporting

Today it is not only environmental pressure groups that seek greater transparency in matters of pollution caused by companies and their products, and about measures taken to reduce or avoid it. It is also the companies' own employees, customers and neighbours that increasingly demand more environmental information. Active communication about environmental issues has grown considerably in importance in recent years – shown not least by the fact that so far more than 1800 German firms have published a free standing environmental report. About 1500 of these reports are environmental statements of production sites that participate in the Environmental Management and Audit System (EMAS) of the European Communities. On a global scale the number of companies or production sites with free standing environmental reports is estimated by the Berlin-based Ecological Economics Research Institute IÖW to amount to about 2500.

Environmental reporting is a new area for entrepreneurial activity. While up to 1990, globally not even 10 companies had presented the public a comprehensive report about their environmental performance, the environmental issue has been giving increasing importance in business reports and special environmental reports since the beginning of the 90's. With the exception of Denmark this is done on a voluntary basis, with no mandatory legal requirement as yet. The Netherlands passed a law on mandatory company environmental reporting in April 1997. From 1998/99 several hundred Dutch companies will have to publish environmental reports on a yearly basis. Since 1995, companies and production sites in the European Union (EU) have, in addition, been given the chance to voluntarily participate in the Environmental Management and Audit Scheme (EMAS) set up by the EU. This requires them to publish a so-called "environmental statement", in which they inform the public about environmental performance and environmental management at a production site.

Company environmental reporting reflects a changed business environment. Four major developments have increasingly influenced business practice since the 80s:

1. the rising need to take ecological requirements into account in business processes and company management,
2. the fact that companies have become more "exposed to the public", hence more subject to the social pressure to legitimise their activities,

3. the increasing influence of the media on social processes and conditions of entrepreneurial activity,
4. the growing importance of information and communication in economic life and their relevance for a company's competitiveness.

The IÖW Study on Effects of Voluntary Environmental Reporting

Research on environmental reporting and communication by companies is as young as its subject. A number of studies and projects concerning company environmental reporting have been carried out since the beginning of the 90s. They focussed on how many companies have published environmental reports and environmental statements, on the requirements placed on reporting on the part of various stakeholders, as well as on the quality of the contents of reports. This research has yielded well-founded results and insights (cf. publications of Clausen, J./Fichter, K.; KPMG; SustainAbility/UNEP).

A matter for further investigation is the benefits to be derived form voluntary environmental reporting. It is obvious that companies produce voluntary environmental reports only if they seem of some use to them. What exactly this use is, in how far it may differ among companies and between sectors, and whether reporting may go along with unconsidered or unexpected positive or negative side effects, remains unclear.

Another issue so far unexplored is the question of how far environmental reporting is done for reasons of competitiveness, and of how far it might affect competitiveness. A potential relationship between environmental reporting and competitiveness will very likely be as ambivalent as the relationship between environmental performance in general and competitiveness (cf. Dyllick 1998). It will most probably fit neither into a simple harmony model such as "environmental reporting benefits a firm's image and business" nor into a simple conflict model such as "transparency will only be of disadvantage for a company".

Against this background, the Ministry of Economy of the German Bundesland Hessen asked the IÖW Berlin to carry out a study on the propagation of voluntary environmental reporting and on its related expenses and benefits. As part of this study, a questionnaire was sent to all 500 German companies and production sites that by September 1996 had published an environmental report and/or an environmental statement according to EMAS (return ratio 46.8%). Based on the respective results, two case studies were conducted, one at 'Hoechst', the pharmaceutical and chemical multinational (worldwide 150 000 employees), and one at 'Hipp', a producer of baby food (750 employees). Hipp is one of the leading producers of baby food in Europe, and is the largest processor of organically grown raw materials on a global scale. The IÖW study was completed in summer 1997. The most essential results and conclusions are represented below.

Results of the IÖW Study

Dealing with a so far largely unexplored field, the empirical research in the context of the mentioned study had, above all, an explorative character. It aimed at ascertaining the effects of a firm's voluntary environmental reporting.

The results of the study corroborate the hypothesis that active environmental reporting supports a company's competitiveness in many ways. Not only has it a positive influence on the company's public image and its relations to social stakeholders, it also contributes to a more clearly-defined market image. In addition, it significantly fosters the quality of management and eco-controlling and increases employee identification with the company.

Of the interviewed companies 95% are willing to publish further environmental reports and statements, which confirms the observation that environmental reporting has so far been distinctly beneficiary to companies. Only 1% of the interviewed companies do not intend to continue this practice while 4% did not answer the question. This result supports the prognosis that company environmental reporting will spread further in coming years.

Environmental Reporting Fosters the Dialogue with Opinion Leaders

From a company point of view, one of the principal effects of environmental reporting is in fostering the dialogue with opinion leaders in environmental associations, media, politics, science and other social groups. Of those interviewed 66% agreed this applied to their own company. The case analyses at Hoechst's and Hipp's show that especially the dialogue with multiplicators such as journalists, top executives, or university professors, as well as with environmental experts in associations and other companies is fostered by environmental reports and statements.

Environmental Reporting Improves Eco-controlling

According to the interviewed companies, one essential benefit of environmental reporting is that better data on environmental performance is made available, which allows improved planning and success in controlling environment performance. Since data and information is to be published in environmental reports, it is collected, processed and evaluated more systematically and more comprehensively than before. The necessity to report stimulates the detection of information and evaluation gaps, thus contributing to a continuous improvement of a company's environmental performance. Hence public reporting fosters the quality of the company's management control.

Environment Protection Becomes more Relevant to the Company Management

Of those interviewed 60% agree with the statement: "Environmental protection has become of greater relevance to the company management, as a consequence of the environmental report and the connected public confession of environmental concern". Hence environmental reporting increases the degree of liability and self-commitment regarding environmental protection. Due to its public character, environmental reporting is necessarily also a matter for top management. Hence top management is faced with a stronger need to tackle the company's environmental impact and environmental performance. Thus, environmental awareness grows in top management.

Different Assessments of the Utility of Reports

As the case analyses at Hoechst's and Hipp's show, the utility of environmental reporting is assessed differently by the various actors inside and outside the company in two respects. First, assessment criteria are different for reporting companies on the one hand and external stakeholders on the other hand, since each party associates different goals with environmental reporting. 'Hoechst' and 'Hipp' want to give credible reports about their environmental protection activities and environmental performance on the basis of data and facts, thereby improving their image and their sales prospects on the market, or fostering dialogue and cooperation on environmental issues. External groups, however, have different priorities. Improved image or sales prospects are irrelevant to them and instead they want transparency and a true description of the environmental situation and performance. Such a description has also to represent ecologically weak points and problematic issues.

So, for example, retail enterprises that buy 'Hipp' products welcome the environmental reports of the baby food producer and assess them as "very good". At the same time, however, they demand reliable standards. Retailers can sensibly use their suppliers' environmental reports as a basis for comparison only if the presented data and information show a (sector-specific) compatibility.

Second, assessment of the utility of environmental reports differ among external stakeholders, for reasons of their heavily diverging information interests. The case of 'Hoechst' may serve as an illustration for this. Environmental experts as found in local citizen initiatives and environmental groups assess the company's environmental reports as largely worthless. For their work, they would need detailed information about individual hazardous substances or about planned facilities. Such information is not contained in Hoechst's environmental reports. Multiplicators, on the other hand, such as environmental journalists, assess the 'Hoechst' environmental report on CD-ROM as very valuable, due to its survey information covering most of its production sites worldwide.

With regard to different information interests of stakeholders, and hence different assessment of the usefulness of a report, our case analyses suggest we can distinguish three main target- or user-groups:

1. the public in general and the firm's employees (as a company-internal public), which are primarily interested in concise, understandable information about subjects or measures that affect them directly,
2. multiplicators or opinion leaders such as journalists, politicians, top executives, analysts or control institutions, which are particularly interested in survey information about the state and development of environmental protection and a company's environmental performance,
3. specialists, like expert representatives of citizen initiatives, environmental groups, authorities, and science, or environmental managers of firms, which are primarily interested in detailed information about individual topics or problematic issues.

Both case analyses show that, independent of their diverging assessments of the utility of reports, companies and various stakeholders consider environmental reporting as useful. Yet their reasons for this as well as their information interests are different. What is generally lacking in reports, however, from a user point of view, is a more standardized structure, a well balanced representation and comparable data.

Environmental Reporting is not of the Same Competitive Relevance to Every Company

The study shows that environmental reporting is not of the same competitive relevance to every company and sector. Publication of environmental information is highly relevant, for example, in the chemical industry, which has to fight a bad public environmental image, or in the food industry, where customers have become increasingly sensitive to environmental and health issues in recent years. In contrast, competitive relevance of active environmental reporting is significantly lower in sectors like mechanical engineering.

Voluntary Environmental Reporting is Done Primarily for Reasons of Competitiveness

Our questionnaires and case studies show that voluntary environmental reporting is done primarily for reasons of competitveness and is performed all the more comprehensively the more competitively relevant it is. For this reason reporting companies can be subdivided into different types of reporters. For example, the big chemical companies, which suffer from a bad environmental image, spend on their reports almost double the sum of big companies of comparable size in other industrial sectors. Companies that produce reports of the type "image problems" spend an average amount of 161 000 DM on an environmental report, whereas companies of other sectors (comparable in size) spend only 88 000 DM per report.

Another illustration of the fact that environmental reporting is primarily done for reasons of competitiveness is given by the following comparison. Despite comparable sizes, companies reporting according to the type "eco-marketing" produce an average number of 3700 copies of their environmental reports, companies reporting according to type "image problems" publish a similarly high number of 2800 copies, whereas for companies publishing environmental statements for the only reason of participating in EMAS this number amounts to only 365 copies.

Reports of type "image problems" are part of society-related competition strategies, by which companies mean to improve their public image and foster their social acceptance in order to secure current markets and products. Reports of type "eco-marketing", on the other hand, are part of differentiation and market development strategies. In both cases, environmental reporting is of competitive relevance. Reports which are published only because this is part of the obligations of EMAS participation, are of very little competitive relevance to their companies, according to the latter's own assessments. Case studies confirm the interview results.

The Competitive Relevance of Environmental Reporting Determines its Quality

Interviews and case analyses support the assumption that the competitive relevance of environmental reporting determines its quality. The degree of reliability and frankness in reporting thus depends essentially on four factors:

1. the demand for information addressed to the company on the part of the market, the public, or political pressure, hence the corresponding "pressure for frankness",
2. the competitive strategy or the compatibility of frank reporting with competition conditions,
3. the company's environmental performance as compared to that of competitors and to legal regulations, and the potential conflict of company-related environmental issues,
4. a company culture that promotes open and self-critical dealing with environmental information.

The quality of reporting results from the interaction of these four factors.

Voluntary Environmental Reporting is Focussing on Good News

It is not surprising that companies intend to present themselves to the outside world in the most favorable light. This applies also to environmental protection activities and environmental performance. Voluntary environmental reporting is not neutral information of external stakeholders but interest-oriented representation of one's own environmental performance. Environmental reporting is a typical form of public relations. The informative contents, reliability and communication style is clearly different from that of publicity and advertising. But nevertheless voluntary environmental reporting is interest-oriented.

Companies want to present data and facts in a credible report about their environmental activities and performance in order to improve their public image, to increase their sales prospects on the market, or to foster dialogue and cooperation regarding environmental issues. External groups, however, have different priorities. Improvement of image or sales prospects are irrelevant to them, instead they want transparency and a true description of the environmental situation. Such a description has also to represent ecologically weak points and problematic issues in a company. So companies focus on the good news, whereas a majority of external stakeholders are interested in problem-oriented reporting.

Largely Lacking Comparability of Environmental Performance

Of all interviewed companies 50% think that their environmental reporting allows a better comparison with their competitors. The publication of environmental data provides the basis for greater transparency and better comparability.

Yet so far existing environmental reports and statements are but a first step in this direction, since sufficiently precise standards, which would ensure comparability, are still lacking. This is not only criticized by environmental associations but also by the trade. Big retail enterprises like the German company Tengelmann plan to use environmental reports for future comparisons among suppliers. But even the guidelines for environmental statements according to EMAS are formulated in too general a way to provide actual comparability.

Ambivalence Between Environmental Reporting and Competitiveness

The study was based on the assumption that the relationship between environmental reporting and competitiveness is as ambivalent as is the relationship between competitiveness and environmental performance in general. It was further assumed that this relationship can neither be described adequately by a simple harmony model nor by a pure conflict model. It has to be noted, though, that the present study could not identify any clear competition-relevant conflicts or disadvantages for companies that actively participate in environmental reporting. Neither did company images suffer from palliative reporting, nor did published environmental data entail competitive disadvantages. The only negative aspects of reporting that might be listed in some cases are the expenses in labour and money to produce the reports.

This result might lead us to conclude that active environmental reporting will always contribute to strengthening competitiveness. Yet such a conclusion would be wrong, for two reasons. First, the present study did not examine how far environmental reporting had negative effects on the competitiveness of those companies that did not publish environmental reports. It may, however, be assumed that this is at least the case for those firms that directly compete with reporting firms. Second, considering the present practical experiences about company environmental reporting, it has to be taken into account that:

- so far, environmental reports are produced voluntarily and environmental statements according to EMAS are published according to not very clearly specified standards. Binding standards and exact legal regulations for company environmental reporting are so far lacking. The great arbitrariness of data presentation allows companies to describe themselves in a favorable light to the external public.
- A majority of the reporting companies thus far show an above-average degree of commitment to environment protection. Actively communicating these advantages allows them to positively distinguish themselves from competitors and to realise "pioneer gains", e.g. in form of an improved image.

Conclusions for Environmental and Economical Politics

Environmental reporting introduces the evaluation of the environmental performance of whole enterprises into market processes. It generates new facts and new information flows. This helps to identify inefficiencies in the value added chain and foster innovative solutions. Criteria for environmental protection and new information modify the selection mechanism in the market. So far, however, those processes are fairly ineffective, since environmental reporting lacks meaning and comparability of environmental performance is still considerably restrained due to lacking standards and regulations concerning information. Besides, there is a lack of professional institutions engaged in purposefully collecting company-related environmental information and processing it so as to be applicable for individual user groups. One first approach in this direction is, for instance, the eco-rating agencies, which evaluate the environmental performance of whole companies and make the results available to banks, fund managers, and institutional investors, etc. A legal framework for an efficient comparison of companies with respect to their environmental performance does not exist so far, with the exception of Denmark.

The need for legal regulation of company environmental reporting may also be concluded from the failure of the idea of 'social reporting' in the 70s and 80s. Main reasons for the "failure of a good idea" then were its tight connection to financial accounting data, its missing user orientation, its lacking legal embodiment, and the use of social reports as a mere public relations instrument.

Environmental Reporting for Promoting Competition in Environmental Performance

The objective of a legal embodiment for company environmental reporting would be to create greater transparency with respect to environmental impact and environmental performance of whole companies. This is supposed to, increase market transparency and market dynamics on the one hand and, on the other hand, decrease informational asymmetries between companies and social stakeholders. Thus, a legal embodiment of company environmental reporting would have to be conceived as a regulation of information, supporting the market mechanism, creating incentives for innovations, and promoting competitively relevant interaction of a company with its social environment.

Such a legal regulation of environmental reporting has to be seen in the context of a new paradigm concerning the relation between environmental policies and competitiveness, which considers the positive dynamic relations between environmental regulation and competitiveness (cf. Porter/van der Linde 1995). Of relevance are not the static cost effects of environmental regulation but the incentives for innovations that ensue from legal standards. Hence environmental standards lead to innovations, which serve for cost reduction and increase resource productivity. Thereby, costs of regulations are balanced, competitiveness is maintained or even increased. Porter and Linde suggest that pollution points at existing inefficiencies in the private sector. External costs of pollution are not included in their observation. Companies produce at a lower level of productivity and do not efficiently use resources.

Environmental reporting proposed here would relate to whole companies, in contrast to that already in existence which is system-related or site-related mandatory reporting. This would introduce qualitatively new aspects into mandatory environmental reporting. So, for instance, environmental reporting of whole companies would also have to consider environmental issues of strategic character (e.g. research and product development or portfolio management of a combination of multinational enterprises). It would furthermore have to focus on the effects of the company's activities during preceding or subsequent stages in product life cycles, as well as on the company's international activities.

Legal Embodiment for Big Companies

Mandatory environmental reporting should be coupled to size and legal form of companies, as is the case with publication duties in financial reporting. This seems sensible for two reasons. First, environmentally relevant facilities and production sites are already subject to numerous individual reporting duties. At present, a serious information gap exists mainly with respect to environmental information that would *transcend* the borders of individual sites or facilities. Second, it can be assumed that big companies are particular polluters and, due to their economic importance, have a particular influence on flows of material and energy, on environmental impacts, and on production and consumption patterns. They are hence of particular eco-political relevance.

Company-related environmental reporting is to be conceived as supplementing already existing reporting duties. Existing facility-related and site-related reporting duties are, in the first place, instruments for the state to meet its control and enforcement tasks. In contrast, company-related reporting would chiefly be designed to provide information to market actors, such as industrial customers, trade, banks, financial analysts, or insurances, as well as to social stakeholders, such as environmental associations, consumer organisations, media, and research institutes. A firm would have to inform about its environmental impact, its environmental performance, and about risks related to its environmental policy. So, company-related environmental reporting is to be understood as a firm's market-related and public-related communication. It would be addressed mainly to experts and multiplicators from the above-mentioned market and social institutions. As has been found in the empirical investigation of the present study, environmental reports are suited to the general public only in a very restricted way. Hence they should primarily be designed as communication with experts and multiplicators rather than as part of the communication with the general public.

Environmental reporting as is proposed here presupposes a company environmental information system (eco-controlling) and an environmental management system. Hence a company legally obliged to produce a company-related environmental report would have to have an environmental management system complying with the requirements of EMAS or the DIN/ISO standard 14001 "environmental management system". The Dutch law on mandatory company-related environmental reporting, passed in 1997, also includes this link between environmental reporting and environmental management system. Similar to the audit of the annual statement of accounts by an auditor, an environmental report should be audited and validated by an independent expert with regard to its completeness, truth, significance, continuity, and comparability.

Thus, company-related environmental reporting is an independent form of company reporting and communication, supplementing not only existing individual reporting duties in the field of environmental protection but also financial reporting that is completed by environment-protective aspects. Each of these three forms of reporting is an independent type of external communication. With regard to the requirements of an integrated company communication, however, all three types should be matched to each other and should complement each other. Company-internal collection and processing of data should also be integrated in a way that avoids double work in the provision of data for different types of reports.

Not only larger companies in the production sector should be obliged to produce a company-related environmental reports, but also firms from the service sector. The business policy of retail enterprises, banks, and insurances also leads directly or indirectly to considerable environmental impacts, and can also contribute to their reduction. In analogy to the publicity law, groups and individual companies from a certain size on (5000 employees and more etc.) could be legally obliged to company-related environmental reporting.

Conclusions for a Company Practice

Four major conclusions can be derived from the present study with respect to company practice and company management:

1. *The need for increased use of society-related strategies to secure and develop markets.* The socio-political environment is increasedly relevant for achieving and maintaining competitive advantages. Hence strategies to secure and develop markets have to be increasingly society-oriented. The "classic" market-oriented competitive strategies of cost leadership, differentiation and concentration on certain points of main effort proposed by M.E. Porter have to be supplemented by and coordinated with society-oriented competitive strategies. Different types of strategies are not mutually exclusive but have to supplement each other in a calculated way.

2. *An integrated and strategic company communication* gains considerable importance, since companies are increasingly publicly exposed. The need for communicative distinction in the market has grown, and requirements regarding company-internal and cooperation-related integrative performance have increased. Public relations, market communication, and internal communication have to be conceived and implemented as integral parts of a comprehensive communication strategy. Public relations are of strategic importance, since they meet with only weakly predefined structures-orientation patterns have still to be developed. This leaves room for proactive action, e.g. for innovative forms of dialogue communication, and to socially position whole companies. Public relations have to contibute to positioning a firm as to market competition, by securing and creating scope for action. In addition, they have to contribute to keeping public welfare in mind and shaping the moral imperatives of business activity.

3. *Strategically shaping the relationship between environmental performance and competitiveness.* Up to the beginning of the 90s, the relation between environmental performance and competitiveness was represented mostly as a cost issue. For some years now, however, emphasis is on the opportunities for distinction on the market and on the potential for cost reduction in environment protection. Neither of these perspectives is wrong, yet closer observation reveals both as one-sided and little differentiated, for they hold only in certain situations. Environmental performance can be of competitive relevance in various ways. It is not only market influences that play a role here. Apart from requirements on the part of customers, banks, and insurances or the strategies of competitors (eco-marketing), social stakeholders (neighbours, environmental associations, media), processes of public opinion formation, and political actors (gov-

ernment, authorities, political parties) have also to be considered. An approach to strategically influencing competition processes has therefore necessarily to be based on a perspective that integrates market, politics, and public. This involves the central question as to what are the (potential, latent, or actual) important problematic environmental issues for a company or for a sector, and how do the different social actors transform these issues into competition-relevant requirements. Social transformation of pollution precedes the emergence of environmental competition fields. A company can only make use of the opportunities resulting from grown environmental requirements if it purposefully considers them in its overall strategy. Doing so, it may offensively deal with environmental requirements and may create appropriate combinations of market-related and society-related competiton strategies for each situation.

4. *Active environmental reporting and environmental communication as integral component of ecological competition strategies.* The relationship between environmental reporting and competitiveness is as ambivalent as is the relationship of environmental performance in general and competitiveness. It is neither adequately described by a simple harmony model such as "environmental reporting benefits a firm's image and business" nor by a simple conflict model such as "transparency will only be of disadvantage for a company". As the present study has shown, active environmental reporting offers numerous opportunities for improving the relation with social actors, for achieving a distinct market position, for motivating staff, and for improving management quality and eco-controlling. Crucial for this is the understanding and use of environmental reporting as an integral part of the company strategy and as an element of environmental competitive strategies. It has to be taken into consideration that environmental communication is shaped by three specific characteristics: first, by the degree of (relative) newness of the environment issue; second, by the conflict potential of environment protection issues, and third, by the triple problem of ascertaining, evaluating and communicating environmental data. These specific aspects influence both company-internal communication, market communication and public relations. In view of the potential conflict of the environmental issue and the triple information problem regarding environmental data, a particular credibility problem arises for external environmental communication. Stable confidence relations and action-shaping images depend essentially on whether a firm's environmental information is reliable and stated facts can be proved. Environmental communication and environmental statements that are meant to contribute to improving a company's image imply direct communication and dialogue and have to consider principles of good environmental reporting.

A convincing environmental communication has to take into consideration that ultimately it is not words but deeds that count in environmental protection. Only those companies that can demonstrate provable achievements as to their environmental performance and pollution reduction will remain credible in the long run.

References

Clausen, J.; Fichter, K. (1993): Vorstudie zum Projekt Umweltberichterstattung, Berlin, Osnabrück.

Clausen, J.; Fichter, K. (1998): Environmental Reports – Environmental Statements – Guidelines on Preparation and Dissemination, INEM/ future e.V., Munich, Wedel.

Company Reporting Ltd. (1996): Corporate Environmental Reporting in the UK, Edingburgh.

DTTI – Deloitte Touche Thomatsu International (1993): Coming Clean, Corporate Environmental Reporting, London.

Dyllick, T.: Ecology and Competitivness, in: Fichter, K.; Clausen, J. (Hg.): Steps to the Sustainable Company (within this volume).

Fichter, K.; Clausen, J. (1994): Wissenschaftlicher Endbericht zum Projekt Umweltberichterstattung, Berlin, Osnabrück.

Fichter, K.; Clausen, J. (1996): Environmental Reports and EMAS-Statements in Germany, Ranking 1995, Berlin

IRRC – Investor Responsibility Research Center (1996): Environmental Reporting, and Third Party Statements, Washington D.C.

KPMG Management Consulting AB (1996): International Survey of Environmental Reporting, Stockholm, Summary Paper presented at press conference September 10th.

KPMG Deutsche Treuhand-Gesellschaft AG (1996): Umweltberichterstattung in Deutschland, eine empirische Untersuchung, Hamburg.

Porter, M.E.; van der Linde, C. (1995): Green and Competitive: Ending the Stalemate, in: Harvard Business Review, p. 120–134.

SustainAbility/UNEP (United Nations Environmental Programme) (1996): Engaging Stakeholders, Vol. 1, The Benchmark Survey, The second progress report on company environmental reporting, London, Paris.

7 Mandatory Public Reporting – A Manufacturing Site's Perspective

Margret Pierce

Introduction

The purpose of this document is to provide a chemical manufacturing site's perspective on mandatory public environmental reporting in the United States.

The topics covered include: reporting requirements, voluntary reporting, examples of environmental improvements, and an analysis of the benefits and the burdens of public reporting.

To put these comments in the proper context, background about the manufacturing site is provided. The Chambers Works site, of the DuPont Company, is located in Deepwater, New Jersey; USA. New Jersey has the most stringent reporting requirements in the United States. The 78-year old site occupies six square kilometers, employs 2400 employees, and ships over 450 tons of product annually. There are over 600 products, including carpet stain resistant chemical, Stainmaster®, alternative refrigerant SUVA®, and intermediates for specialty fibers, Kevlar® and Nomex®. The site houses a corporate research and development facility and the USA's largest industrial waste water treatment plant.

Due to its size and complexity, the site is the largest submitter of mandatory environmental forms in the state of New Jersey. In addition to the reporting required by regulations, Chambers Works participates in voluntary waste reduction programs and issues a safety, health and environmental progress report to the public annually.

Mandatory Public Reporting Requirements

In 1987, the United States began requiring certain industries to publicly report environmental releases under the "Toxic Release Inventory". Over the years, the list of chemicals grew, as did the required data. Currently, DuPont Chambers Works is required to provide data about 65 of the 600 chemicals and categories. This effort requires the equivalent of one person working full time for six months. The majority of the work involves gathering production records, yields and inventory; determining quantities recycled and disposed; and compiling facility-wide totals for each of the 65 chemicals (see Fig 1 for the site's historical Toxic Release Inventory data). In addition to the Toxic Release Inventory required by the Federal government, the state of New Jersey requires additional materials accounting data such as the starting inventory, quantities produced or brought on site, consumed on site, shipped off-site and the ending inventory. This data is also required annually and consumes the equivalent of an additional full time person working for six months. Again, facility-wide totals are reported for each chemical.

The latest addition to New Jersey's mandatory public reporting is Pollution Prevention Planning. The state requires facilities who use large quantities of hazardous chemicals to go through a specific planning process and submit a publicly available commitment to hazardous chemical use and non-product output reductions. (Non-product output is any material that is not shipped as or in product. This includes waste prior to treatment, fugitive and stack emissions, and out-of-process recycled material.) In this program, planning and reporting are mandatory, but reduction goals are voluntary. Facilities have the option of submitting reductions of 0%. This flexibility allows the manufacturers to decide how best to use their limited resources. The state could have required a fixed reduction for each process. This would have made it much more difficult for companies to optimize overall reductions.

Pollution prevention plans must be done every five years, and annual progress reports are submitted. The initial plan required the equivalent of over two full time people for a year (program coordination, documenting the plan's requirements, "brainstorming" sessions which included engineers, operators, chemists and mechanics).

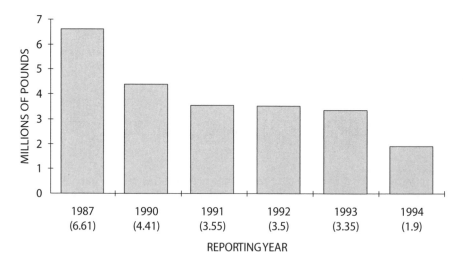

Fig. 1.   Chambers works toxic release inventory releases

Pollution prevention planning consisted of quantifying the amounts used, released, and non-product output for each chemical, for each effected process. In order to demonstrate a financial incentive for pollution prevention, the plan required facilities to tabulate all the costs associated with using or producing hazardous chemicals and wastes. The processes that made up 90% of the usage or non-product output were then "targeted". Targeting required detailed pollution prevention analysis to be conducted. After pollution prevention options were developed and prioritized, reduction goals were determined. Chambers Works submitted reduction goals of 16% and 53% for use and non-product output, respectively.

Table 1.   Summary of reporting requirements

	Toxic Release Inventory		Pollution Prevention
	Federal	State	State
Reporting Requirements	Quantities (reported as ranges)	• Starting Inventory	• Facility-wide non-product output reduction goal for each chemical
	• Amount released to air, water, land	• Quantity produced, brought on site	• Facility-wide use reducti-on goal for each chemical
	• Amount transferred off-site	• Quantity consumed on site	• Process descriptions
	• Amount recycled (on & off-site)	• Quantity shipped off-site as or in product	• For each "targeted" process:
	• Amount treated	• Ending Inventory	non product-output re-duction goals per unit of product; and implemen-tation schedule;
	• Amount burned for energy (on & off-site)	• Quantity recycled on-site	use reduction goals per unit of product and sche-dule for implementation
	• Amount disposed	• Quantity destroyed through on-site treatment	
	• Production ratio	• Quantity released to air and water	
	• Source Reduction Activities	• Quantity treated on-site	
Staff Requirements	~ 8 employees for a total of > 0.5 person-years	~ 8 employees for a total of 0.5 person-years	Initial plan >2 full time equivalent people; Updates: ~ 0.25 person-years
Cost	$50 M Chambers Works > $1 MM DuPont Corporation	$50 M Chambers Works	Initial Plan: $200 M Updates: $25 M

Internal Programs to Drive Environmental Improvements

Environmental improvements, in addition to being driven by regulations, are the result of many voluntary and internal DuPont programs.

The site participates in the Responsible Care® initiative. (Responsible Care® is the United States' chemical industries' voluntary program to bring industry performance in line with public expectations.) Each year, the site does a self-audit to gauge progress against the Responsible Care® "codes". One of the six codes in particular, "Pollution Prevention", provides a good benchmark for determining the completeness of your site's pollution prevention efforts.

Three examples of the 14 management practices that comprise the Pollution Prevention Code follow (taken from "Responsible Care® Guide", DuPont 1994):

Management Practice 2: A quantitative inventory at each facility of waste generated and released to the air, water and land, measured or estimated at the point of generation or release.

Benchmark
A site has an inventory if the following information is available: measurements, calculations, records and report which quantify solid, liquid, and gaseous wastes generated by the operations of the site and the subsequent quantities and compositions of the discharges to the environment and to treatment, disposal and recovery facilities.

Management Practice 3: Evaluation, sufficient to assist in establishing reduction priorities, of the potential impact of releases on the environment and the health and safety of employees and the public.

Benchmark
The evaluation of impacts of releases must consider not only the environmental impact, but also the health and safety impact to employees and local public. This process is evergreen, and should be used to assist in establishing waste and air emissions reduction priorities, consistent with other corporate and business safety, health and environmental goals.

Management Practice 6: Ongoing reduction of wastes and releases, giving preference first to source reduction, second to recycle/reuse, and third to treatment.

Benchmark
Progress in reducing wastes should be measurable.

As another example of a voluntary program, Chambers Works participated in the United States Environmental Protection Agency's "33/50" Program which was aimed at reducing the 1988 release of 17 high volume toxic chemicals by 33% by the year 1992 and 50% by the year 1995. The site reduced releases of these chemicals by 78% (see Fig. 2.).

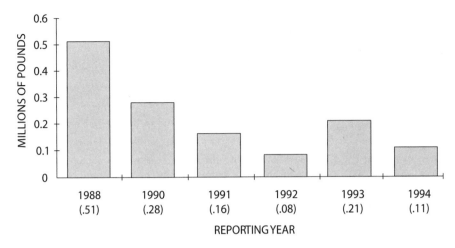

Fig. 2. US environmental protection agency's "33/50" voluntary program

There are also many internal DuPont programs that help reduce wastes:

- The corporation has a policy of "The Goal is Zero Injuries, Wastes and Emissions"
- The Corporate Environmental Excellence Program issues prestigious awards to employees who make significant environmental improvements. (This award has recently been expanded to recognize Safety and Health Excellence.)
- Accidental spills and releases are reported, assigned a "severity" category and tracked against a site metric. A 50% improvement goal is set each year.
- The site issues an annual Safety, Health and Environmental progress report. Toxic Release Inventory releases are reported, as are other corporate metrics such as airborne carcinogens and toxic air emissions. A similar report is issued for the entire DuPont corporation.
- Monetary awards are given to employees when they suggest environmental improvement ideas that are put in place.
- Mechanics, operators, chemists and engineers participate in pollution prevention brainstorming sessions. These various perspectives truly complement one another, and solutions are developed because of the exchange of ideas and knowledge.

Examples of Pollution Prevention Activities

Most of the pollution prevention progress that has occurred at the site has been driven by a need to reduce costs or improve utility. Although public reporting encourages Chambers Works to find ways to reduce wastes, projects in DuPont are not authorized strictly on the basis that they will reduce reportable chemicals; they must be accompanied by an appropriate return on investment. Some recent examples are as follows:

(Note: The following examples are taken from the "DuPont Chambers Works Waste Minimization Project", May 1993.)

Case 1: CAC Process

One process at the Chambers Works site produces several chlorinated aromatic compounds (CAC) in separate product "campaigns". The processing equipment includes reactors, distillation columns, and storage tanks. These must be cleaned between campaigns to prevent old product residue from contaminating the new product. Until recently, the cleaning was accomplished by flushing the process equipment with a solvent, producing about 600,000 lbs. of solvent waste for incineration every year.

In 1990, an operator serving on a CAC waste reduction team suggested that the process equipment should simply be drained instead of flushed. This idea led to the installation of drainage valves at low elevations on the process equipment. The operator, with the help of two mechanics, designed and built a "mobile collection vehicle" essentially a 55-gallon drum secured to a wheeled carriage. Now at the conclusion of each product campaign, workers wheel the collection vehicle to each drainage valve, collecting the product residue, and dumping it into a storage tank where it is held for reprocessing during a future campaign. The solvent flush has been eliminated entirely.

The amount of residue within the process equipment after draining is negligible, and does not compromise the quality of the next batch. As an added benefit, the new drainage scheme takes less time than the old solvent flush, which means less downtime between product campaigns.

This low-tech solution to a large waste disposal problem cost just $10 000 to implement. For this, DuPont will realize a saving of Net Present Value of $2.2 million over ten years. The CAC experience demonstrates the importance of including line workers on waste reduction teams. Technical people tend to arrive at technical solutions to the problem of waste generation. But the operators and mechanics who operate the process every day can often identify simpler, cheaper and more effective solutions.

Case 2: SAC Process

One process area at the Chambers Works site makes several specialty aromatic compounds (SAC). The process consists of a reaction step which produces the SAC, and a distillation step which purifies it. A waste stream from this process contains large amounts of SAC product entrapped within a heavy tar which forms as a byproduct of the reaction step. The value of the lost SAC and the cost of incinerating the waste stream provided a strong incentive for waste reduction.

The SAC reaction step produces two types of tar. One type, called "thermal" tar, is an inevitable consequence of the high-temperature reaction. The other type, "acid" tar, forms when one of the raw materials used in the reaction is too acidic. The ratio of thermal tar to acid tar in the SAC waste is unknown because there is no practical way to distinguish between them.

The raw material responsible for generating the acid tar is produced elsewhere on the Chambers Works site. In the past, the acid content of the material has fluctuated greatly. This used to frustrate tar reduction efforts at the SAC process by making it impossible to measure the success of those efforts.

In 1990, the raw material process began using a pH meter to monitor pH levels and keep acidity low. This resulted in an immediate tar reduction downstream at the SAC process. Moreover, reliable knowledge about raw material pH enabled people at the SAC area to implement at last waste reduction solutions of their own. Their efforts to date have reduced the SAC waste stream by almost 60%. More reductions are planned.

Installing the pH meter at the raw materials process cost $50 000. The savings from lower incineration costs and increased product recovery will total a Net Present Value of almost $8 million over ten years.

In several instances, the Chambers Works project demonstrated that a waste reduction program must encompass all the activities that contribute to producing the waste. People at the SAC area had long attempted to reduce the waste generated by their process. But not until they expanded their efforts to include the "upstream" raw materials process were they able to achieve a breakthrough.

Case 3: Polymer Vessel Washout

One process area at the Chambers Works site makes several grades of a certain polymer. The process equipment includes a large vessel which must be cleaned periodically to maintain product quality. During processing, polymer accumulates on the interior vessel walls, agitator blades, and baffles. Cleanups are complicated by the vessel's construction, which renders opening the vessel difficult and time consuming.

Until recently, the vessel was cleaned by flushing with a flammable solvent. This produced a waste stream of about 40 000 lbs. of solvent for incineration each year. But in 1991, a process improvement team eliminated the solvent wash and implemented a system that uses a high-pressure water jet.

In the new system, a specially designed nozzle is connected to a high-pressure water source and inserted through the bottom flange. A stream of water at 10 000 psi (pounds per square inch) blasts the residual polymer from the interior surfaces. For safety reasons, the entire cleaning process is operated remotely, and no high-pressure spray escapes the vessel. The non-hazardous polymer residue is at present collected and landfilled. But the area marketing group is now at work investigating commercial uses for the polymer residue.

Implementation of the new cleaning system cost $125 000 dollars. The savings which result from the elimination of the solvent waste will total Net Present Value of about $2.7 million over 10 years.

As with many of the Chambers Works case studies, the polymers story illustrates the importance of economic drivers to waste minimization. The replacement of the solvent wash originated not from a dedicated waste reduction effort, but from an overall process improvement program that had waste reduction as just one of its goals. Other goals of the program included quality improvement, shorter cycle times, and reduced operating costs. The new vessel cleaning system was implemented because it met all of these goals.

Benefits of Mandatory Public Reporting

There is truth in the statement "what gets measured gets attention". In the absence of mandatory reporting, companies have no external driver for toxic release reduction. Although some will recognize the benefits of reducing wastes, others will not. The Toxic Release Inventory has resulted in release reductions over the years it has been in place.

The presence of information such as chemicals and quantities used provides local emergency planning groups with facts that they need in order to do an effective job in preparing for an emergency. In addition, the community has a right to know what is being released into their surroundings. Toxic Release reporting gives them the opportunity to obtain that information.

When public attention is drawn to the toxic release information through the local newspapers, it provides a company with an opportunity to take credit for progress made in reducing releases and it increases the interactions and openness between the manufacturer and those living nearby.

The more detailed material balance data that is required by the state of New Jersey (i.e., amount used and efficiency of use) provides public interest groups and environmentalist groups with data that could allow them to focus on companies with less "green" performance.

For companies who otherwise would not quantify their wastes at the source, New Jersey forces them to do so. This knowledge could encourage them to develop pollution prevention solutions.

As far as the benefits derived from mandatory pollution prevention planning and reporting, they are similar to some mentioned previously. Companies who commit to large reductions of waste benefit from the positive publicity they receive. Companies, like DuPont, who are going to do pollution prevention planning anyway, might as well benefit by sharing the commitments publicly. New Jersey requires companies to assess the costs associated with producing wastes. For some companies, this could be an "eye-opener" that drives pollution prevention.

Ideally, waste should be reduced at the source. This is the method that results in the greatest environmental improvement, reduces the risk of worker exposure and transportation incidents, and if economical solutions can be found, it saves the most money. New Jersey's pollution prevention planning focuses on reducing waste at its source.

Lastly, the State took a very positive step towards balancing the economy and the environment by making the reduction goals voluntary. Companies therefore have the flexibility to use their scarce resources to maximize pollution prevention results instead of trying to achieve a fixed reduction (i.e., 10%) across all processes.

Burdens of Mandatory Public Reporting

For a facility as large and complex as Chambers Works, Toxic Release Inventory reporting requires a tremendous amount of resources. The resources are environmental engineers, production assistants, clerks, managers who review the results, and corporate professionals who compile the data for the corporate performance. In a global marketplace, this puts the United States (and other countries with reporting requirements) at a competitive disadvantage.

A lot of negative publicity occurs for companies "at the top of the list". There are several reasons that this negative publicity may be misdirected.

1. There are companies in other industries that would be "at the top of the list", but they are not required to report (utilities, commercial waste treaters).
2. Major chemical producers can be at the top the list because of the magnitude of their production, not necessarily because their processes are inefficient.
3. The inventory is based on quantity, not toxicity. Thus, little attention is brought to producers of smaller streams of highly toxic wastes.
4. Non-point sources are not included so pollution that occurs from automobile exhaust, agricultural run-off, and consumer waste, although they are as significant in quantity and environmental harm, are not addressed.

DuPont's safety record is 10 times better than the chemical industry, and the chemical industry is 2 times better than manufacturing in general. The Toxic Release Inventory is aimed not only at reducing releases to the environment, it is also aimed at use reduction. A company like DuPont, with an excellent safety record has shown that hazardous chemicals can be used and manufactured safely and responsibly.

The materials accounting portion of the New Jersey requirements may create a strategic advantage for competitors. Competitors have access to capital investment dollars from public announcements when a company builds a new plant or expands substantially. This information combined with material balances and production data, could give a competitor knowledge of another's cost position which it could use to its advantage in making competitive quotes for contract businesses. With knowledge of pounds of product per investment dollar, the competitor can estimate the selling prices required in the future to be economically viable. This helps a competitor in deciding long term pricing which in turn helps in making their investment decision.

To assess the effectiveness of the Toxic Release Inventory, it must be pointed out that:

1. The list of chemicals changes continuously, so it is very difficult to track progress.
2. By the time the data is made public, the data is well over a year old.
3. In New Jersey, duplication of work results from the Federal and the State governments requesting similar data, but on different forms.
4. Environmental groups and public interest groups have the staffs, resources and energy to understand the Toxic Release Inventory, but the general public does not.

The burdens of New Jersey's mandatory Pollution Prevention Planning will be discussed. The most burdensome aspect of the Plan is that it requires companies to do a lot of paperwork even if they have already assessed their processes and implemented solutions. For instance, Chambers Works was starting up a hydrochloric acid (HCl) recovery unit at the time the initial plan was due. Even though the pollution prevention solution was already installed, the Plan required that we document the process flow sheet, the costs associated with having HCl as a waste, the options considered, and the technical and economic feasibility of each option. Clearly, this was a waste of resources. DuPont Chambers Works has been implementing source reduction for many years. The site contends that the processes that contribute to ninety percent of the non-product output of the plant are well understood. It was burdensome to have to go through the exercise of tabulating every process' non-product output, releases, uses and costs prior to selecting the processes to further analyze for pollution prevention.

Similar to the Toxic Release Inventory, companies are driven to reduce the high volume waste streams that are usually not the most toxic (they are often dilute acid streams that can be easily treated in a waste water treatment plant and rendered harmless). In addition, out-of-process recycling is not counted as pollution prevention, only closed loop recycling may be counted.

Lastly, a burden, especially in uncertain business climates, is forecasting five-year reduction goals that usually require capital investment. In DuPont, capital expenditures are forecast for long range strategic planning, but commitments for capital spending usually occur on an annual basis. Therefore, a plant is required to submit a reduction commitment well in advance of receiving a timely capital spending commitment.

8 Environmental Benchmarking in Italy

Matteo Bartolomeo, Federica Ranghieri

Environmental Benchmarking

Benchmarking is the process of comparing and measuring an organisation's business process and performance against a given standard.

It is a tool currently used in several business areas and particularly in quality management. The objective of benchmarking is the promotion of process or product improvement by the identification of a recognised standard and of the related actions required.

The insights gained from benchmarking provide an organisation with a foundation for building operational plans to meet and surpass the standard and promote an overall awareness of environmental improvement opportunities. Benchmarking can promote thinking that generates improvement breakthroughs and leads to greater awareness of the need for long term planning to address current and emerging environmental issues (GEMI 1994a).

Despite the wide use of benchmarking techniques in quality, marketing, finance, technology innovation, the expression *environmental benchmarking* is still a vague concept; the implementation of benchmarking techniques to the environmental area should be appropriately studied, conceptualised and tested.

Environmental benchmarking is an environmental management tool that can provide a substantial contribution to the improvement of environmental performance by facilitating the identification of the gap between company performance and a given performance.

Any process or business activity can be a candidate for environmental benchmarking and particularly the loop Planning-Doing-Checking-Acting can be usefully assisted by this instrument. Similarly, many activities foreseen by EMAS regulation or ISO 14000 (especially initial review, objective identification, programme definition, audits) can be improved by environmental benchmarking tools.

From the literature and through practice it is possible to subdivide environmental benchmarking in four main categories:

- *Internal benchmarking.* This evaluation is carried out by company management via questionnaire and audits and has the aim of improving the loop policy-targets-programmes-results. Internal benchmarking can also help company managers in identifying SWOT and therefore in improving economic efficiency of the company. There are many examples of this kind of evaluation. The GEMI (Global Environmental Management Initiative 1994a) has developed a guide for self assessment of environmental performance with particular reference to management practices. The GEMI methodology uses as a benchmark the ICC Business Charter for Sustainable Development and has the aim of assisting companies in evaluating the gap from Charter principles. The EMAS regulation, the BS 7750 or the new ISO 14000 series requirements can also be considered a useful benchmark for internal performance evaluation.

- *Best in class benchmarking.* This has the aim of identifying best practice in environmental management. Such an exercise frequently involves companies from different sectors distinguishing themselves for the effectiveness of their management systems. This kind of benchmarking exercise is generally sponsored by a pool of companies willing to share information and suggestions for the improvement of their management system; in the environmental area, AT&T and Intel have for example worked together for the improvement of environmental management practices.

- *Competitive benchmarking.* Since business competition is more and more based on environmental performance, companies' interest for understanding the position of their competitors in an eco-efficiency scale has grown rapidly and induced some of them to develop competitive benchmarking tools to be applied to the environmental area. This evaluation is mostly undertaken by strategic consultants using confidential information in addition to environmental reports and other publicly available information.

- *Sector benchmarking.* Sector benchmarking is normally undertaken by industry associations with the aim of collaborating with authorities and the stimulation of business competitiveness by assessing the average performance of the sector and differences among individual companies. While at present the application of these techniques to the environmental area has a narrow objective of publishing sector reports, a broader a more challenging application of sector benchmarking could be greatly beneficial for stimulating firms' continuous improvement of environmental performance.

Table 1 shows scope, objectives, audience and instruments of a wide range of performance evaluation (benchmarking): internal benchmarking, best in class benchmarking, competitive benchmarking and sector benchmarking.

Benchmarking techniques are becoming quite popular in the US but are still at the age of infancy in Europe. Large corporations, which are more and more familiar with performance measurement have been among the first to apply benchmarking tools to the area of environmental management.

The use of these techniques in Italy is very limited to a few well known examples. On the other hand there are a certain number of competitive environmental benchmarking cases which are just partly known since performers normally prefer to keep information and findings very confidential.

Table 1.   Different meanings for environmental benchmarking

Kind of evaluation	Scope	Objectives	Audience	Instruments
Internal benchmarking	The company	Improve competitiveness; Stimulate continuous improvement; Improve economic efficiency; Undertake SWOT Analysis; Allocate environmental cost; Find effective employee rewarding systems.	Company managers	Questionnaires and audits
Best in class benchmarking	Small number of companies	Identify best environmental management practices among selected companies.	Company managers, Industry associations	Joint benchmarking projects, literature
Competitive benchmarking	Competitors	Identify performance, objectives, strategies and programmes of competitors; Identify best practices.	Company managers	Literature, Strategic consultants Analysis, Environmental reports, Environmental statements
Sector benchmarking	Specific sector Industry branch	Identify sector strategies and programmes; Get the knowledge to negotiate with authorities; Define parameters for eco-labelling; Disseminate information on best practices; Define training packages.	Industry association, Governments, Industrial district authority	Joint projects via questionnaires and bulletins, Environmental reports, Environmental statements

Despite the small number of benchmarking applications to environmental management, the Italian situation shows interesting examples which are briefly illustrated below: an ongoing project in sector benchmarking involving the ceramic
tiles industry and a competitive benchmarking analysis performed by Eni.

Another interesting application of environmental benchmarking techniques has
been performed in 1994 by Italian branches of Landis & Gyr, IBM Semea, SGS-
Thomson, Texas Instruments and 3M[1].

---

[1] The project has been co-ordinated by Consorzio Autofaber, a joint venture between IBM, Italtel and
Milano Ricerche.

This project, called *Benchmarking for Environment*, had the aim of identifying best attitudes and practices on environmental management through the identification of 50 key issues. The 50 elements where selected among two broad categories, management systems – ranging from environmental policies to product stewardship – and specific programmes and actions including performance related to all media emissions and to consumption of raw materials and energy. The confidentiality on findings and on the benchmarking process unfortunately do not allow an in depth description of this project which has been remarkably effective for the participating companies.

Environmental Benchmarking for the Ceramic Tiles Sector

The Ceramic Tile Sector: Economic Importance and Environmental Problems

Italy is the world largest producer of ceramic tiles for covering walls and floor with a production of over 550 million m² of ceramics per year. This production, which accounts for about the 25% of world production, is mainly concentrated in a relatively small industrial district located in the north of Italy (80% of Italian production in an area of about 50 km²).

Such a concentration of production activity has generated and still generates large environmental impact especially due to water discharges and road transportation of raw materials and products.

In 1991, as a consequence of the decrease in environmental quality and of several environmental crises, the district was named a *high environmental risk area* by the Italian Ministry of the Environment, stressing the problem of long term sustainability of industrial activity.

On the other hand, the ceramic tile industry, with its 260 sites and related businesses, is by far the most important employer with more than 21 000 people working in the sector.

The economic and social importance of the ceramic sector and the severe threat for the environment has induced local authorities, the industry association and individual companies to undertake important initiatives to promote more environmentally conscious production activities.

As part of these efforts, the transformation of furnaces to gas power, a wider use of best available technologies and logistic optimisation have played an important role in decreasing the negative environmental impact of ceramic tile production. Today, as a consequence of stakeholder's pressures, companies have partly changed their behaviour and modified production technology (we will call these measures, *hardware innovations*); the sector is now among the most active in Italy in terms of environmental protection and the *greening* of companies is becoming key to maintaining an evident competitive advantage.

While technological improvements are now widespread within the sector, environmental management efforts, we could name them *software* measures, are still weak. There are several reasons for this and particular importance can be attributed to the small size of companies in this sector, and therefore to the lack of management systems in general.

Environmental management systems, audits, communication with stakeholders, green design are not considered, by most of the companies, as effective efforts to comply with environmental legislation nor a means to improve economic efficiency of the company.

Companies' strategies are therefore still largely dominated by *hardware* measures which have had a very beneficial influence on local environmental conditions. On the other hand – since the marginal environmental improvement is now decreasing together with the cost effectiveness of new environmental technology the adoption of management measures could promote a substantial increase in competitiveness of individual companies and, more generally, of the sector.

Companies are therefore in a *wait and see* position in which the behaviour of neighbours (namely competitors) can strongly and rapidly influence the company attitude. In order to promote the adoption of environmental management measures, the ceramic tile industry association (Assopiastrelle) has launched a number of initiatives: a project on the definition of sector's guidelines for the participation to EMAS scheme, a pilot implementation of environmental management systems in some companies and a major *environmental benchmarking* project, jointly carried out by the Fondazione Eni Enrico Mattei and Centro Ceramico of Bologna.

Structure of the Environmental Benchmarking Project

The project on environmental benchmarking has been launched in order to stimulate competition between ceramic tile companies on the basis of environmental performance. As pointed out before, firms have a *wait and see* attitude that could be positively turned into good environmental management practices if elements of novelty are introduced in some companies and become known within the whole sector.

Information and comparison among companies could therefore open a new field of competition and have positive feed backs on the quality of the environment and on competitiveness of companies.

The project is structured into different phases:

1. Identification of Information to be Gathered
The information that has been identified should comply first of all with criteria such as significance, availability, comparability. Moreover, the information to be generated by this project refers both to efforts and to results. This dichotomy, which complies with the structure of the draft ISO 14031 on environmental performance

evaluation, is due to the need of linking the economic and managerial efforts with the environmental performance.

The questionnaire is therefore divided into two sections: the first addressing environmental management systems issues – such as the existence of an environmental policy, of an initial environmental review, the attribution of clear responsibilities, the use of tools to communicate with the public and the economic efforts in terms of operating and investment expenditures. The latter referring to the physical flows of inputs, products and all media emissions.

The whole project focuses on environmental issues but also on quality management, product performance, energy efficiency, technology and health and safety (with a specific set of questions for each issue). The inclusion of these issues in the project, due to purposes other than improvement of environmental performance, will have beneficial feed backs on the environmental part of it as well.

2. Sampling and Data Collection

The whole sector in Italy accounts for about 290 companies (260 in the district): all of them are involved in the project. The decision to address the whole sector has been very controversial from a financial point of view. It is justified by the need to generate enough information to make useful comparisons within a sector which, even if made of similar businesses, cannot be considered fully homogeneous in terms of acquisition of raw materials, production process, product mix and integration of production phases.

Such a comprehensive data collection phase will be performed through half-a-day face to face interviews that will represent a comprehensive and self consistent firm check up.

3. Aggregation, Comparison, Analysis

The aggregation of data, starting at the end of May 1997, will take into account the above mentioned differences within the sector. These differences, not particularly relevant to benchmark environmental management attitudes, can on the other hand determine higher or lower environmental performance which cannot be attributed to company mismanagement of environmental issues. A sensitivity test will therefore allow a division of the the sector into sub-sectors in order to properly identify best, average and poorest performers and disclose to each company the gap from these performance levels.

The project will not lead to an identification of an aggregated performance indicator but of a set of areas and indicators relevant to describe company efforts and environmental performance.

The large amount of information related to different areas (not only the environmental one) will allow a better understanding of the relationship between general and environmental management attitudes, technology adopted and environmental performance. It will show for example the correlation between the adoption of quality systems and environmental management systems, or the link between overall technology and environmental performance.

4. Eco-efficiency Bulletin Preparation

In order to stimulate an environmental competitive behaviour at firm level, the project foresees the delivery of an *Eco-Efficiency Bulletin*, a brief report in which the environmental performance of an individual company is ranked with those of competitors and compared with the average, with the best and worse performer.

The *Eco-Efficiency Bulletin* will also safeguard the confidentiality of the information provided by each company, a basic ingredient for the success of the entire project: an individual company is able identify its own competitive position in relation to benchmarks.

Furthermore the *Eco-Efficiency Bulletin*, represents an incentive for companies to disclose data on environmental issues and, together with the individual check-up, will constitute a tangible and direct benefit for companies participating to the project.

5. Sector Report Preparation

The project foresees the preparation of a sector environmental report – prepared by the Fondazione Eni Enrico Mattei in co-operation with Centro Ceramico di Bologna and published by Assopiastrelle – this will summarise the efforts and environmental performance of the whole sector. The sector report will represent an effective instrument to communicate with external stakeholders, as demonstrated by previously published environmental reports, issued for example by the Chemical Industry Association in the UK and other countries, or the American Petroleum Institute.

A confidential report will be delivered to the industry association describing mean, standard deviation and suggested actions to address specific issues in order to stimulate the environmental performance of the entire sector and the competitive attitude of individual firms.

The confidential report will also enable Assopiastrelle to dialogue with local and national regulators, and possibly to negotiate agreements on the basis of a comprehensive understanding of sector current and future performance.

6. Training and Dissemination of Findings

Training courses will be organised with the objective of sharing information on best practices and to promote the competition of companies on the basis of their environmental performance.

Sources of Difficulty

Since the project is still at the beginning, commenting on project findings, both in terms of effectiveness for the industry association and in the promotion of improvement in environmental performance in member companies, is still premature.

Nevertheless it is possible to consider some sources of difficulties that require, in this but also in other sector environmental benchmarking projects, prompt and appropriate answers.

First of all the *reliability of information collected.* Since companies have an incentive to act as free riders, an early involvement in the project and the delivery of tangible and direct benefits (like the check up and the bulletin) play an important role and should reduce the risk of collecting meaningless data.

Secondly, the *selection of environmental performance indicators* should be based on the general principles "simple, clear, significant and comprehensive" but at the same time allow a sophisticated analysis of data for delivering the confidential report to the industry association.

Thirdly, the *comparability of companies* is in practice not as obvious as is it in theory. While the ceramic tiles sector can be considered a very homogenous industry; some differences can affect the reliability of results. It therefore becomes crucial to solve the trade-off between creating comparable sub-sectors while keeping a significant sample of firms in each sub sector.

Competitive Benchmarking in the Petrochemical Sector

Eni, the large Italy based and state owned conglomerate operating in the oil, gas, chemical and energy sector, decided in 1996 to set up a competitive environmental benchmarking project to evaluate environmental performances both for external communication purpose – whether to publish the first Eni Environmental Report or not – and for internal management purpose – to test economic efficiency in environmental protection, to verify improvements and to set targets.

For this analysis, carried out jointly by Fondazione Eni Enrico Mattei and Eni staff, several competing firms have been selected, namely: Akzo Nobel, BP, British Gas, Dong, Eni holding (and separately Eni Group companies: Enichem, Snam, Agip Petroli), Exxon, Gas de France, Montecatini, Neste, Polimeri Europa, Shell, Statoil. Two main steps have been taken in this environmental competitive benchmarking analysis:

- the first step consisted of a comparison of environmental management practices in ten companies selected from the ones mentioned above. This information has been obtained with an analysis of companies' environmental reports published in 1994, 1995 and 1996, and referring to activity in 1993, 1994 and 1995.
- the second step consisted of a comparison of environmental performance, related, for example, to the quantity of resources consumed, energy efficiency, quantity and kind of emissions.

The first step has been defined as a *qualitative environmental benchmarking* and the second step as a *quantitative environmental benchmarking* analysis. The first and the second steps together offer a comprehensive view of companies' policies, management systems and their results recorded in different years.

The study gathered and assembled information and data on business practices and performance indicators related to the following ten focus areas:

1. Policies, Plans and Procedures
2. Targets
3. Environmental Management Systems
4. Monitoring and Documenting Performance
5. Health and Safety Indicators
6. Resource Management
7. Environmental Performance Indicators
8. Emissions Control and Reduction
9. Environmental Expenditures
10. Incidents and Remediations.

The First Step

The first part of the analysis has focused on qualitative aspects and on the identification of the existence of issues that have been quantified and compared in the second step.

Table 2 illustrates for example the result of the analysis of environmental policy and targets, based on different questions. To evaluate the existence of the environmental policy three questions have been formulated:

- is there a serious commitment?
- are there any basic principles or set priorities?
- is this commitment reflected in paragraphs or in other parts of the report (CEO's address, few general words, etc.)?

Table 2. Policies and targets (*Yes* means that the company provides specific information)

Company	Existence of Environmental Policy Statement	Type of Environmental Policy	Existence of Targets in the Policy Statement
Company A	yes	HSE	yes
Company B	yes	HSE	no
Company C	yes	HSE	no
Company D	yes	E	yes
Company E	yes	E	no
Company F	yes	E	no
Company G	yes	HSE	yes
Company H	yes	HSE	yes
Company I	yes	HSE	no
Company L	no	–	no

Table 3.   Environmental management system

Company	well-defined policy	management support	training	risk management	accounting support	commu- nication	inter- action
Company A	yes	yes	yes	yes	no	yes	n.a.
Company B	yes	yes	yes	no	yes	yes	n.a.
Company C	yes	yes	yes	yes	no	yes	n.a.
Company D	yes	yes	yes	yes	no	yes	n.a.
Company E	yes	yes	yes	no	yes	yes	n.a.
Company F	yes	yes	no	yes	no	yes	n.a.
Company G	yes	yes	yes	no	no	yes	n.a.
Company H	yes	yes	yes	yes	no	yes	n.a.
Company I	yes	yes	no	yes	no	yes	n.a.
Company L	no	no	no	no	no	yes	n.a.

In order to evaluate the implementation of environmental management systems, a precious set of basic issues have been isolated including a well-defined policy, management support, resources, training, emergency preparedness, regulations, accounting support, communication areas, interaction with other functional areas (i.e. engineering, production maintenance, legal departments, etc.).

At the first stage, the analysis allows also to identify the kind of quantitative information (performance indicators) companies disclose in their environmental reports. An in-depth analysis of environmental performance has been carried out in the second stage. At the second stage this competitive environmental benchmarking analysis enables a more technical comparison between companies' environmental performance.

In terms of air emissions for example, a number of indicators – VOC, NO_x, SO_x, CO, CO_2, CFC – have been selected but not always used by all the companies in their environmental reports. Similarly for discharges to water the indicators identified have been: Total Discharges on Water, BOD Indicator, COD Indicator, Oil, Chemicals, Incidental Spills.

The Second Step

Sixteen companies environmental reports published in 1994, 1995 and 1996, with data and information about their 1993, 1994 and 1995 activities have been compared, focusing on environmental performance, namely resources consumption: energy, water and raw materials; emissions to air, water discharges and waste; emissions and consumption related targets; and environmental expenditures.

Companies use a number of specific measurements in their environmental reports. These include measures that would be expected based on legal concerns (quite similar across sectors) and measures used to meet internal corporate goals. This variety in measurement and in disclosure can help explain the difficulties of the team of analysts involved in this project.

This quantitative benchmarking has been developed on three consequent levels:

- corporate level
- business area level
- site level.

At a corporate level Akzo Nobel, BP, British Gas, Dong, Eni, Exxon, Gas de France, Montecatini, Neste, Shell and Statoil have been compared. To benchmark them, performance indicators, were needed. The most common performance indicators to measure over time companies environmental performances are operational indicators, which enable the evaluation of the eco-efficiency in the use and consumption of resources and emissions from activities, relating those to the companies productions.

At a corporate level, even if considering companies within the same sector, it is impossible to determine comparable physical output basically due to differences in a business's sominant product (a 80% chemical company's productions and emissions should be very different from a 80% oil refining company's or from a 80% gas company's productions and emissions). Thus, the calculus of the performance indicators related the consumption and emissions to the companies value added instead of production levels.

By using value added as denominator, the sample has been reduced to only six companies, the only ones disclosing enough information to obtain a rough index

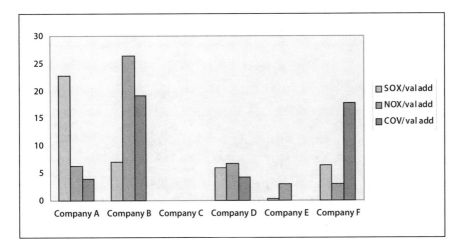

Fig. 1.   Emission to Air: tons of pollutants per million US$ of value added

representing their value added. Anyway, the comparison results were not significant because of differences in disclosure methodologies and lack of data and comparable information.

At a business area level the team selected four main sectors: Chemicals, Oil & Gas Exploration and Production, Gas and Oil Refining.

To measure over time companies environmental performances and benchmark them, standard operational indicators were used: resources consumption and emissions related to the companies production levels.

Enichem, BP Chemicals, Neste Chemicals, Shell Chemicals and Polimeri Europa were compared, but not all the comparisons were significant: too many differences in business activity and products (even within the chemical sector, petro-chemical and pharmaceutical companies, for example, present quite different production processes and related emissions), and on disclosure methodologies.

For the Exploration and Production sector five companies have been selected. To measure and benchmark them, the team needed to take into account the organisational frameworks: some activities were world-wide, some other national (Agip, Shell UK and Statoil); activities offshore and onshore presented different predominance across the sector; and the production of gas versus oil is quite variable.

The Gas sector's environmental reports published the smallest quantity of data, mostly because of the estimated lower impact on environment. It was very difficult to compare and benchmark companies in this sector because of the lack of data: the only available comparisons were about emission to air: NO_x/production, SO_x/ production and methane/production.

The best results were obtained studying the oil refining sector. Agip Petroli, BP, Exxon, Neste Oil, Shell Oil, Statoil were compared over three years 1993–95. The analysis within this sector was conducted at a site level: six refineries were compared. The team selected refineries considering similarity in processes, products,

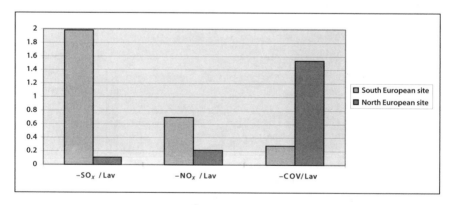

Fig. 2.   Emission to air: oil refining sector, site level analysis (attention: SO_x/LAV should be considered as SO_x/crude oil as input)

facilities and plants. The comparison results were interesting: not only environmental strategies and assets emerged, but also differences from internal practises and different regulation. For example, North European sites usually present SO_x/ production lower that Southern European sites, partly because of stronger regulation in the environmental field.

The first result of this environmental benchmarking project has been the publication of the 1995 Eni Environmental Report. Considering the data and information available, Eni decided to publish in 1996 its first Environmental Report, a document studied and prepared only to communicate with stakeholders. The second result was a rough evaluation of effectiveness in environmental protection.

As briefly explained, the project has met several obstacles mainly due to the differences in kind of information disclosed and in indicators used. Nevertheless it is expected in the future a process of convergence in environmental disclosure and hoped a voluntary standardisation of indicators will be used at least at industry level.

Eni will continue this comparison and use benchmarking tools to set environmental targets and to compare its performance with those of competitors.

References

Azzone, G.; Manzini, R. (1994): Measuring Strategic Environmental Performance, Business Strategy and the Environment.

Business in the Environment (1993): The measure of commitment, London.

European Green Table (The) (1993): Environmental Performance Indicators in Industry, Oslo.

Fitzgerald, C. (1992): Selecting measures for corporate environmental quality: examples from TQEM companies, Total Quality Environmental Management.

GEMI (Global Environmental Management Initiative) (1994a): Environmental Self Assessment Program.

GEMI (Global Environmental Management Initiative) (1994b): Benchmarking: the Primer.

James, P.; Bennett, M. (1994): Environmental Related Performance Measurement in Business, Ashridge Management Research Group, England.

James, P.; Bennett, M. (1994): Financial Dimension of Environmental Performance, Ashridge Management Research Group, England.

JPC and SBC, report of the workshop Environmental Benchmarking, 1997

Krafter, B. (1992): Pollution prevention benchmarking: AT&T and Intel work together with the best, Total Quality Environmental Management.

NMPC (1993): Environmental performance index, Syracuse, NY.

Wells, R.P.; Hochman, M.N.; Hochman, S.D.; O'Connell, P.A. (1992): Measuring environmental success, Total Quality Environmental Management.

9 Full Cost Accounting as a Tool for Decision-Making at Ontario Hydro

Helen Howes, Ali Khan, Corinne Boone, Takis Plagiannakos, Barb Reuber

Abstract

The paper describes Ontario Hydro's approach to full cost accounting (FCA) and its experience in using it as a tool for integrating environmental considerations into its business decisions. It also identifies barriers and opportunities for implementing the concept and illustrates, through examples, ways in which FCA has helped in better understanding the potential environmental costs and liabilities associated with Ontario Hydro's activities and in reducing the impact and cost of those activities now and in the future.

Introduction

Ontario Hydro, serving the Province of Ontario, Canada, is one of the largest electric utilities in North America in terms of installed generating capacity. Total in-service system capacity is approximately 29 000 megawatts, transmitted across 29 000 kilometers of transmission lines and 109 000 kilometers of distribution lines. Its customers include 306 municipal electric utilities, which in turn, serve more than 2 800 000 customers, and Ontario Hydro Retail which serves almost 1 000 000 retail customers, including 103 large industrial customers.

Ontario Hydro owns and operates 69 hydroelectric stations, 5 nuclear stations and 6 fossil fueled stations. Ontario Hydro's electricity generation in 1996 was 55% nuclear, 26% hydroelectric, 13% fossil and 6% other. Total revenue in 1996 was $8.9 billion on an asset base of $40 billion. The company currently employes approximately 21 000 people.

In 1993, the Task Force on Sustainable Energy Development (SED), commissioned by the then Chairman Maurice Strong, identified full cost accounting (FCA) as a key component of Ontario Hydro's strategy for achieving SED. In April 1995, Ontario Hydro's Board of Directors approved SED Policy and Principles which provided guidelines for implementing SED in the organization. Two of those guiding principles related to a need for developing a framework for using full cost accounting in decision-making:

- Ontario Hydro will integrate environmental and social factors into its planning, decision-making, and business practices.
- Ontario Hydro will practice eco-efficiency, that is continuously add value to products and services, while constantly reducing energy use, material use, pollution and waste.

This paper will focus on Ontario Hydro's approach to full cost accounting (FCA) and its experience in using it as a tool for integrating environmental considerations into its business decision-making processes. It also identifies barriers and opportunities for implementing the concept and illustrates through examples, ways in which FCA has helped in better understanding potential environmental costs and liabilities associated with Ontario Hydro's activities and in reducing the impact and cost of those activities now and in the future.

Full Cost Accounting and Sustainable Development

Full cost accounting essentially entails collecting and considering all costs associated with a certain activity, process or product (through out its entire life cycle) in decision-making.

There are numerous definitions of FCA and it is therefore important to define what FCA means to Ontario Hydro. At Ontario Hydro, FCA is defined as a means by which environmental considerations can be integrated into business decisions to:

- better understand and allocate its internal environmental costs;
- better define, quantify and, where possible, monetize the external environmental impacts of its activities; and
- integrate environmental impact and cost information (qualitative, quantitative and/or monetized) into planning and decision-making.

Ontario Hydro's approach to FCA (Fig. 1) has been focused from the start on influencing the decision-making processes at various levels in the organization, rather than using it as a cost accounting or reporting tool. Therefore, FCA has been promoted within the organization as a tool for factoring *internal* and *external* environmental costs into the decision-making process, in addition to the financial evaluation, socio-economic analysis, and risk assessment practices that take into account other factors like price, reliability, customer service, financial soundness, uncertainty and risk. Effective integration of economic, environmental and social factors in decision-making would result in:

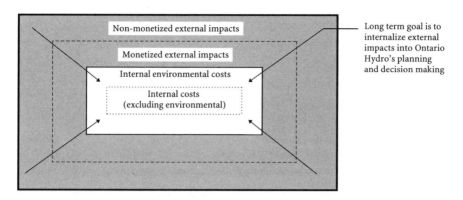

Fig. 1.   The concept of full cost accounting

- cost savings through waste reduction and pollution prevention initiatives,
- early identification and avoidance of future environmental costs and liabilities,
- improved environmental performance and competitiveness,
- transition towards a more sustainable energy future, and
- development of new "green" products and business opportunities.

Figure 1 illustrates the concept of full cost accounting in more detail.

Internal Environmental Costs

The inner core, labeled as internal costs, are the utility's cost of doing business and are often referred to as private costs. These include traditional business costs such as, material, labor, fuel and depreciation. These may also include some internal environmental costs such as, costs associated with meeting environmental regulations or corporate environmental standards (e.g., operating, maintenance and depreciation cost of emissions control equipment, premium paid for use of low sulfur coal as fuel, and on purchase of other "environmentally friendly" products). There are other less tangible (or hidden or indirect) costs such as: costs of contingent liability or production loss from station deratings due to exceedance of regulatory limits, that should also be included in this category. These costs are often not identified, allocated or reported separately within the traditional accounting systems. By ignoring these costs, though, a business unit may fail to account for the true costs of its products and services and thus may make inappropriate business decisions.

Ontario Hydro's approach to FCA differs from other approaches, in that, it includes environmental externalities (societal costs) in its assessments and hence deserve more explanation.

External Environmental Costs (Externalities)

External impacts (externalities) are the impacts on human health and the environment (natural and socio-economic) resulting from the production or use of a product/service, which are not reflected in its cost or price. An external cost (benefit) represents the monetized value of an externality. An externality exists when the following two conditions are met:

- an activity by one agent causes a loss/gain of welfare to another agent, and
- the loss of welfare is uncompensated (Pearce and Turner 1990).

An example of an externality would be human health effects associated with air emissions from fossil-fired electricity generation stations. An example of an externality which has been internalized would be compensation to a farmer for crop damages that could result from ozone.

It is important to note that what Ontario Hydro is trying to incorporate into the decision-making process are the "residual" impacts, or in other words, the environmental and social impacts which remain after all regulations have been met and mitigation and compensation have been undertaken. However, even after Ontario Hydro has complied with all the environmental regulations to control air emissions from its generating stations, there are still residual air emissions that can potentially cause damages to the environment and human health. Ignoring these impacts and costs underestimates the environmental impacts of activities and the resulting costs to society, and may result in inappropriate resource allocation decisions being made. Ontario Hydro believes that explicit consideration of these impacts and costs in decision-making will lead to more sustainable decisions, improved environmental quality, and lower societal costs.

Ontario Hydro has used the Damage Function approach to estimating the externalities associated with its activities, rather than the Cost of Control approach. The Damage Function approach considers:

- site-specific environmental and health data;
- how emissions/effluents are transported, dispersed, or chemically transformed using;
- environmental modeling techniques;
- how receptors (i.e., people, buildings, fish, forests) are affected by these emissions/effluents; and
- the monetary value of these physical impacts.

Ontario Hydro has developed preliminary external cost estimates for operating its fossil stations, and for fuel extraction through to decommissioning for its nuclear generating stations. In addition, it has assessed the externalities associated with the hydroelectric stations and other renewable energy technologies like wind turbines, solar, and biomass located in Ontario. In situations where the external impacts cannot be monetized qualitative assessments have been done.

Applications of FCA

1. Use of FCA in Investment Decisions

All investment proposals that require senior management approval go through a Business Case Analysis (BCA) process. One of the components of the BCA is a discussion of the "SED Implications" of the proposal in light of the SED Decision Criteria. The Decision Criteria is a form of SED screen that is consistent with Ontario Hydro's FCA approach, and focuses on five key elements of an investment decision:

- maximizing resource use and energy efficiency;
- reducing environmental damage/impacts;
- avoiding/improving social impacts;
- increasing the use of renewable technologies; and
- identifying financial gains attributable to SD initiatives.

The analysis of the environmental impacts of the preferred investment option, and the alternatives should consider:

- full life-cycle, where possible, but at a minimum, must consider impacts associated with the design, construction/production, operation/use, decommissioning, and disposal;
- expected damage to ecosystems, communities, and human health, and not be based on the ability to meet existing or proposed environmental regulations;
- identification and evaluation of the potential positive and negative environmental impacts of the alternatives, including impacts which are common among the alternatives;
- quantification and monetization of the potential impacts, where possible, but as a minimum requirement, a qualitative assessment of the impacts; and
- trade-offs which were made in selecting the preferred alternative.

The role of staff specialists reviewing the BCA is to provide advice to the decision-makers (generally the President and CEO) based on an independent review of the SED implications of the business case. Since September 1994 when the SED Decision Criteria were implemented, over 20 BCAs have been reviewed. A majority of these addressed the criteria appropriately and were recommended for senior management approval. In some cases, the SED implications analysis proved valuable in identifying alternatives and developing win-win solutions. The analysis also exposed financial staff, who typically prepare the BCAs, to sustainable development issues and helped fulfill the "systematic reporting" element of environmental due diligence.

In one particular case, a proposed investment decision for a $24 million transmission line refurbishment program, the SED implications were:

- reduction in energy loss in transmission lines through the use of energy efficiency conductors;

- million annual increase in revenues through the re-use and re-cycling of removed line components;
- initiation of a program to improve the biodiversity of rights-of-way by restoring and replacing natural habitats; and
- provision of employment and economic benefits to local communities.

This investment decision was approved and the program is now operational.

2. FCA in Corporate Integrated Resource Planning

The Corporate Integrated Resource Planning (CIRP) process was initiated in the summer of 1994 and was completed the following year. The purpose of the CIRP was to provide strategic advice to the President and CEO on resource allocation decisions for the 1996 business planning cycle. A range of demand-side management and generation supply options were combined into seven plans and evaluated on the basis of their ability to fulfill the following objectives:

- provide competitively-priced energy services valued by customers;
- improve environmental performance and make more efficient use of resources;
- enhance social and economic benefits in Ontario; and
- enhance the financial, operational and human resource viability of Ontario Hydro.

These objectives were used to develop the criteria by which the plans were assessed and evaluated. One of the assessments was an environmental assessment. The environmental assessment was scoped to include the biophysical environment only; impacts on human health and the social environment were considered in separate assessments. The primary criterion established for the environmental assessment was to minimize damages to the environment. The measures used were:

- incremental land use (ha);
- crop damage ($) resulting from ground-level ozone;
- damage to exterior of buildings ($) due to acid gas and
- particulate matter;
- acidic deposition (mg/sq. m.) on sensitive watersheds;
- waste generated (Gg by type of waste);
- water flow modifications due to new hydroelectric developments (water flow ratio);
- impacts of once-through cooling on littoral zones (index based on number, flow, capacity,
- and mode of cooling water system);
- greenhouse gas (GHG) emissions (Tg and Tg/TWh);
- radioactive waste in storage (Mg); and
- consumption of non-renewable resources (i.e., coal, uranium, gas, limestone) (Mg).

The environmental assessment for the CIRP represented an advancement from such assessments Ontario Hydro has performed in the past, in two areas. First, the assessment was performed on an environmental damage basis, which was consistent with the FCA Corporate Guideline to use the damage function approach for monetization of environmental externalities. Second, some environmental damages were mapped on a watershed basis and compared to indicators of ecosystem vulnerability. Multi-criteria Analysis (MCA) was used to assist in making trade-offs among the eleven environmental measures in order to select the most important environmental indicators for evaluating the CIRP plans. The maps of the indicators of ecosystem vulnerability were used together with the maps of environmental damages to assist with this process.

3. FCA for Eco-efficiency Improvement

In late 1996, an eco-efficiency study was undertaken at one of Ontario Hydro's fossil generating station. Eco-efficiency is defined as a process of continuously adding value to the product (electricity generation) while constantly reducing resource use and generation of waste throughout its life-cycle. In order to accomplish eco-efficiency improvements, it is essential to know the resource input (material, fuel, water, energy) and output (energy, by-products, waste, and releases to the environment) from the station, and their associated costs. This requires taking inventory of all environmental activities and operating equipment required to meet, or exceed, environmental regulations and corporate standards, during station operations.

Total internal environmental costs were estimated to be approximately 21% of the station's total operating and fuel costs. Emissions monitoring and control, collection and disposal of ash, coal pile management, environmental regulatory reporting and support were identified as major contributors to these costs. Most of these costs were either fixed or were required to run the station well within regulatory limits. However, some eco-efficiency opportunities were identified in the area of heat rate optimization, energy efficiency, by-product sales, equipment service life extension, and resource and waste management. At the time of writing this paper, work on the project is on-going.

4. Life Cycle Review of Light Vehicles

A life cycle review (LCR) of light vehicles was undertaken to provide input to the Commodity Management Team at Ontario Hydro to assist in their procurement decisions. The objective of the LCR was to compare the life cycle impacts of alternative vehicle fuel cycles: gasoline, diesel, natural gas, propane, and alcohol; with particular focus on life cycle emissions, efficiency and cost. The assessment discounted vehicles that were not commercially available. Although the focus was on environmental impacts and cost, technical and social factors were considered to a lesser degree.

Result of the LCR indicated that the life cycle costs of vehicles that travel long distances (>35 000 km/year, generally used by meter readers or operations staff) are the lowest for natural gas and propane fuels. These options offer significant cost reduction and emissions reduction when compared to gasoline-fuelled vehicles. In addition, there are strategic advantages to using alternatively-fuelled vehicles in Ontario Hydro's fleet in the areas of sustainable development, public perception, culture change, and positioning for the future.

5. Managing Internal Environmental Costs

A pilot project was undertaken at one of Ontario Hydro's thirteen electricity distribution utilities to test and demonstrate the value in knowing the internal environmental costs associated with a business activity. Internal environmental costs and liabilities form a significant portion of the utility's total costs. By knowing exactly what they are, how much they are, and which activities are causing them, the utility can better manage its environmental costs and liabilities. This requires searching for innovative ways of achieving the same or better environmental performance at reduced cost through better planning (i.e., environment built in from the start, not an add-on) and more efficient processes, with minimum waste generation.

Total environmental cost for the utility was estimated to be approximately 8% of its total operating costs. Several opportunities for cost reduction, revenue generation, cost avoidance, and environmental improvement were identified along with recommendations for achieving them. Some of these initiatives are already underway, while others may require further technical and economic evaluation prior to their implementation. The initiatives identified were cost effective and if implemented, were expected to result in a net income improvement of 5–15% to the utility's bottom-line.

6. Potential Application of Environmental Externality Estimates

Previously, the focus of externality research within Ontario Hydro was to inform significant resource decisions, likely through a Corporate Integrated Resource Plan process. However, with a diminished likelihood of central resource planning, externality impacts and costs will likely now inform a larger number of smaller resource decisions, primarily at the Business Unit level.

As the external impacts and costs are quantified and monetized, the external impacts and costs are expected to be used to:

- support Business Case Analyses;
- assess the environmental dispatching of the system;
- contribute to decisions about retiring or rehabilitating existing stations;
- evaluate benefits and costs of new proposed environmental regulations;
- evaluate benefits and costs of additional pollution control equipment; and
- evaluate environmental externalities associated with imports and exports of electricity.

7. Research on External Impacts and Costs

Over the last few years, Ontario Hydro has undertaken research to identify, quantify and, where possible, monetize the externalities associated with the generation and delivery of electricity in the province of Ontario.

Preliminary external cost estimates have been produced for the operation of all the coal-fired fossil stations. These estimates include impacts of air emissions on human health, crops, lakes, buildings, etc. In addition, further research is under way, in cooperation with the Federal Government of Canada, to improve the monetized values of the health effects associated with air pollution.

External cost estimates for the nuclear stations have been produced for the full fuel-cycle (i.e., from uranium fuel extraction and mining to electricity generation and decommissioning). In addition, a study has been completed in assessing the perception of risk associated with the operation of the nuclear stations.

Most recently, externality cost assessments have been completed for our hydro-electric stations and other renewable energy technologies like wind turbines, solar PV, biomass, bio-gas and micro-hydro.

In the future, research will be carried out to assess externalities associated with transmission lines as part of the Environmental Assessment process. In addition, research will be initiated to develop an integrated externality assessment framework for the routine quantification, monetization and updating of the externality estimates associated with fossil generation. The research on nuclear externalities will focus on accident risks, long range impacts (e.g., arising from high-level waste disposal), and public perception of accident and radiation risk.

Lessons Learned from Ontario Hydro's Approach and Implementation of FCA

During the development, testing and roll-out of its approach to FCA, clear barriers emerged:

- FCA is not yet mainstream thinking, it is seen as being in a developmental phase;
- the concept, definitions, and terminology are still not clear;
- it is seen as accounting and reporting framework, rather than decision-making framework; and
- there is legal concern about being held liable for the external costs identified.

However, there are opportunities for overcoming these barriers. Below are some of the lessons learned by Ontario Hydro to date, in its effort to develop and implement FCA.

1. FCA Must be Positioned as an Approach Which Makes "Good Business Sense"

In the same way that environmental issues are often not seen as business issues, FCA is not seen as a business issue. Steps must be taken to demonstrate the benefits of understanding the environmental impacts and costs (internal and external) associated with business activities to show potential for reductions in costs now and in future environmental costs and liabilities. If this is done, then FCA will be considered to make "good business sense".

Case studies and projects where FCA has been applied and have contributed to a better business decision, provide concrete examples that may facilitate change in acceptance.

2. It is Important to Implement FCA as Part of the Corporation's Environmental Management System (EMS)

In this regard, it is important to develop corporate FCA Guidelines or policies/ strategies and link them to the EMS. It is then best to let materiality guidelines determine degree of implementation.

3. It is Important to Establish a Champion and Rationale for FCA

In this regard, FCA implementation:

- needs a senior manager in the organization to champion its value and use for business decisions;
- should be developed and implemented as part of a larger context; for Ontario Hydro, sustainable development is that context.; and
- should be clearly linked to investment decisions.

4. It is Critical to Clearly Define What FCA Means for Your Corporation

For Ontario Hydro, FCA is a means to facilitate integration of environmental considerations as a component in business decisions. It incorporates internal AND external impacts (qualitative and quantitative) and costs/benefits into business decisions.

For Ontario Hydro, FCA is NOT:

- the only decision-making process, but rather an element that is integrated into existing corporate decision frameworks;
- full blown monetization of ALL internal and external impacts and costs, impacts that cannot be monetized should also be considered, but qualitatively;
- an "Accounting System" (i.e., Financial Management System); and
- full cost pricing.

5. FCA is Multi-disciplinary by its Very Nature and Requires a "Team" Approach

The successful development and implementation of FCA requires a team approach with input from a variety of professionals in the organization such as: scientists and planners, environmental economists and management and financial accountants.

Ontario Hydro stresses that it is essential to build bridges between environmental and financial staff (i.e., managerial accountants) in the organization. Many of the capital investment decisions are made within the financial area of the organization. If investment proposals are to be considered on more than just private costs, then there must be communication and collaboration between the financial and environmental decision-makers in the organization.

6. Developing and Implementing FCA is a Gradual Process. It Will not Happen Overnight

The process of developing and implementing FCA is data intensive and time consuming. It is best to focus on those areas where it is possible to exert the most influence and obtain positive results. It is also important to emphasis potential competitiveness benefits.

Training and communication are KEY to drive the right behaviours of analysts and managers. Analysts and managers need to understand how FCA can help in arriving at better decisions.

Ontario Hydro has also found that engaging its business units in externality research increases awareness of the environmental implications of activities. In turn, this can highlight areas where actions in the present could reduce, and possibly even eliminate, potentially significant future environmental liabilities.

7. Environmental Leadership is Important to Customers

Ontario Hydro's public attitude research indicates that environmental quality is a priority for customers. Customers also have high expectations that Ontario Hydro will demonstrate leadership in managing the environmental effects of its operations. Ontario Hydro takes these stewardship responsibilities very seriously.

Conclusions

Ontario Hydro firmly believes that FCA provides a useful tool by which it can better understand and manage its environmental impacts and costs, both now and in the future. However, as Ontario Hydro moves to a more competitive environment, there will be a number of challenges. For example, how can Ontario Hydro achieve continuous improvement in environmental performance without significantly increasing the resources currently dedicated to environmental protection? To gain this advantage, current approaches and expenditures need to be examined and alternatives sought at a lower cost or higher performance level. This approach requires good environmental impact and cost data as well as ongoing monitoring and verification in such areas as pollution prevention, air emissions reductions, and management of conventional and toxic wastes.

Ontario Hydro supports the view that the companies which will be setting the competitive standard in the future will be those companies that see environmental requirements and issues as business opportunities, and not just added costs. For that reason, Ontario Hydro's initiatives will continue to focus on strengthening the relationships among economic/financial, environmental, and human resource management performance to enhance competitiveness. FCA will be a critical tool in the this process.

References

Boone, C.; Howes, H. (1996): "Accounting for the Environment," CMA Magazine.

Boone, C.; Howes, H.; Reuber, B.(1996): "Ontario Hydro's Experience in Linking Sustainable Development, Full Cost Accounting and Environmental Assessment," Electricity, Health, and the Environment: Comparative Assessment In Support of Decision-Making, International Atomic Energy Agency.

Ontario Hydro (1996): Internal Environmental Cost Review of Southwest Hydro, Internal Report.

Ontario Hydro (1996): Life Cycle Assessment of Cars, Minivans & Light Pick-up Internal Report.

U.S. Environmental Protection Agency (1996): Environmental Accounting Case Studies: Full Cost Accounting for Decision-Making at Ontario Hydro.

10 From an Individual Company's Environmental Management to Substance Chain Management

Kathrin Ankele

Limits to Company-Confined Activities

The increasing complexity of production systems, with corresponding international division of labour, and the growing relevance of responsibility extending over the whole product life cycle, point at limits to an individual company's efforts in environmental performance. If environmental programmes are analyzed which companies publish in the context of their participation in the Eco Management and Audit Scheme (EMAS) or the international environmental management standard ISO 14001, it turns out that for a while already preceding and succeeding stages of a company's activities are also taken into account. A multitude of targets and measures in environmental programmes already relate to suppliers and customers. With complex product chains, product-related environmental protection depends considerably on the ecological quality of preceding and succeeding stages. But companies will have to include those stages more systematically and more actively than they do today so as to fully exploit the inherent optimization potentials. The situation is best exemplified by a trading company that does not produce itself but purchases all merchandise from suppliers: its assortment policy, its logistics, the disposal of its products after use etc. determine its environmental performance, which it can only improve in cooperation with other companies (suppliers, forwarding agents, disposal institutions ...). The respective optimization potentials are not confined to the field of ecology, they also refer to the economic area, and, in the sense of sustainability, have also social implications. But the latter aspects will not be discussed here in more detail.

Entrepreneurial cooperation along a product chain with the aim of ecological optimization has been introduced into the environmental debate as a new concept under the notion of 'substance chain management' by the inquiry commission "Protection of Humans and Environment" of the 12th German Bundestag (see Fig. 1).

This concept takes into account different trends in current environmental policy: on the one hand a changing orientation, turning from a media-based perspective to one that is based on material flows, on the other hand the increasing importance of self-initiatives on the part of industry and consumers.

Practical experiences with substance chain management and equivalent activities in recent years have shown that usually an initiative is started in an individual

---

Inquiry Commission: "Substance Chain Management ...

... is the target-oriented, responsible, holistic and efficient influencing of material flows or material systems, with set targets coming from the ecological and economic fields, and considering the social aspects.

Targets are developed at company level, at the level of the chain of actors involved in a substance chain, or at the governmental level."

Fig. 1.   The definition of 'substance chain management' on the part of the inquiry commission
Source: Enquête-Kommission "Schutz des Menschen und der Umwelt" (1993)

company, a major part of preparatory activities takes place there, and further companies along the product chain are included only at that stage.

This insight confirms the relevance of company-internal information instruments and management methods also for the solution of problems transcending the scope of the individual company.

Starting Point: Company-Internal Environmental Management

An existing company-internal environmental management system provides a promising basis for substance chain management projects. The present text will show the connection between company-internal environmental management systems and substance chain management.

An environmental management system according to EMAS serves to continuously improve a company's environmental protection, to warrant compliance with all environment-relevant laws and regulations, and to inform the public about the company's environmental performance. Figure 2 lists the elements of the environmental management system according to EMAS, which will be shortly explained below:

A company formulates its environmental policy, which describes its long-term environmental protection strategy and its target orientation. This policy is to cover important aspects mentioned in EMAS like, e.g., control and reduction of environmental impacts, product planning, and prevention of accidents. The initial review identifies weak points with respect to legal compliance, to the environmental management system (or of the organization of environmental protection if there is no management system yet) and with respect to environmental impacts. In the next step, the environmental management system is implemented. It orients by the needs of the respective company and is supposed to ensure continuous improvement of

Fig. 2. Elements of the Eco
Management and Audit Scheme
(EMAS)

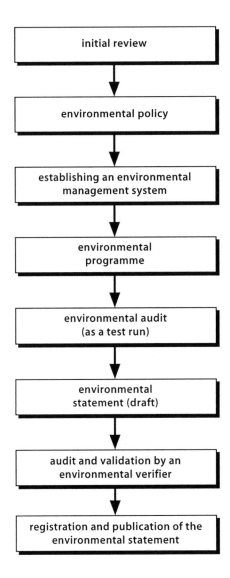

the company's environmental performance. Therefore, its tasks and processes have
to be embedded into a continuous process of eco-controlling. Weak points detected
in the initial review are translated into targets of the environmental programme,
which has to be reformulated for each audit cycle (usually a period of 3 years) and
which determines concrete quantitative goals, measures for their attainment and
temporal, financial and personal resources. Then follows the environmental audit,
which can be performed by internal or external auditors and which focusses on
the effectiveness of the environmental management system. Afterwards, the company produces an environmental statement according to EMAS requirements, and

an environmental verifier is appointed. He or she verifies the validity of the environmental statement and of a random sample of its individual statements, by visiting the company, checking documents, discussing with employees etc. At last, provided there are no objections on the part of the environmental authorities, the environmental statement is assessed as valid, i.e. validated, and the site is finally registered both nationally and internationally.

Three aspects are to be emphasized in a company's environmental management, which are of relevance also in substance chain management:

- information,
- organization,
- communication.

The following chapters will individually discuss these aspects and describe where starting points for substance chain management can be found.

The Site-Related Eco-Balance as a Data Base

EMAS or ISO 14001 do not explicitly require a site-related eco-balance, nevertheless it is a very well suited information instrument to perform the required systematic gathering and evaluating of data concerning environmental impacts.

An eco-balance ascertains, for a given period of time, all material and energetic inputs into a company as well as outputs leaving it (see Fig. 3). Product-related or process-related representation of a company's material and energy flows, combined

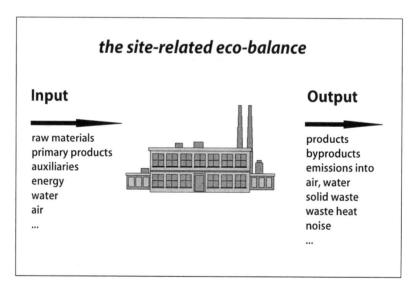

Fig. 3.   Diagram of a site-related eco-balance. *Source*: own depiction, IÖW (1996)

with environmental cost accounting that also orients by material and energy flows, allow to identify ecological and economic weak points in a company's processes. Consideration of raw materials, auxiliaries, and primary products on the one hand and of ways of product disposal on the other hand extend the scope of action beyond company borders. This provides a starting point for substance chain management along the product chain.

Process of Continuous Improvement

In the first place, an eco-balance is a static information instrument representing material and energy flows that have occurred in a company up to a certain fixed day. Yet a company's environmental performance can be improved continuously only if targets and tasks are explicitly defined and organizationally integrated. This requires a controlling process including the elements of planning, regulation and control. A tried and tested procedure in that sense is eco-controlling. It allows to continuously formulate ecological targets, identify weak points to be overcome and optimization potentials to be exploited, according to the environmental programme. Since these activities are integrated into everyday business, they become a matter of course and so do not gradually lose their effectiveness after implementation.

The Transition to Substance Chain Management

There is a fluid boundary between site-related environmental management and substance chain management, which Fig. 4 below will explain diagrammatically. Substance chain management transcends the borders of the individual company and comprehends preceding and succeeding stages of a different vertical range. Substance chain management in this context is not an end in itself, it is a sensible concept only if it is of use for the participating actors. Objectives and expectations of cooperating partners often differ. And in any case, cooperation has to be economically realizable to justify the additional expense, since cooperative solutions often show increased complexity.

Basically, the same three aspects are of importance in substance chain management as in a company's environmental management:

- information,
- organization,
- communication.

But complexity with respect to these aspects is higher, as several companies and maybe further actors are involved (like stakeholders, governmental actors etc.). For reasons of secrecy, information is not exchanged as readily as among different

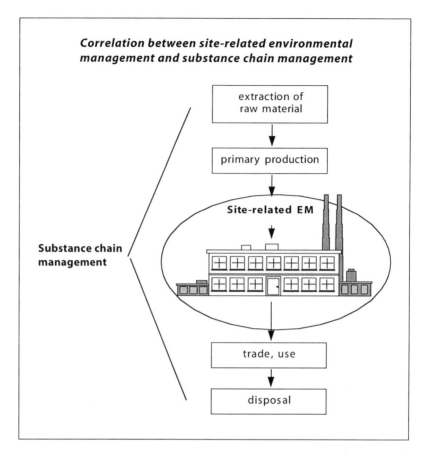

Fig. 4. Correlation between site-related environmental management and substance chain management. *Source*: own depiction, IÖW (1997)

company-internal departments. Goals related to a substance chain management project may differ among companies and may hence provoke conflicts.

Like the similar company-internal process, a substance chain management project needs technical, procedural, and power promoters to initiate and advance the process. Technical promoters bring in the required know-how to make qualified decisions regarding product alternatives. Procedural promoters know how to initiate and advance cooperation, hence contribute with their organizational know-how. Power promoters are those that are able, due to their market power, power to convince, etc., to achieve cooperation of other companies respectively those that should absolutely participate in a specific cooperation along a product chain.

Figure 5 below shows what phases are of importance in a substance chain management project – with not each phase occurring in each project. Analogy with the above-listed steps of the eco-controlling process is conspicuous. We may draw a

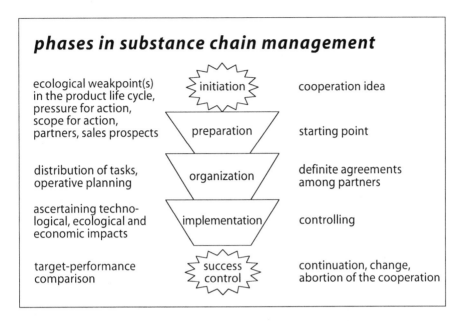

Fig. 5.   Phases in substance chain management. *Source*: own depiction, IÖW (1996)

somewhat simplified analogous picture, showing a weak point at the beginning, which means need for action. Then follows the search for alternatives. This search leads to the formulation of targets and definite agreements on how to attain them. If that process is organizationally integrated in each company concerned, this will allow to implement the respective measures and to continuously monitor their impacts. Finally, the success of the described measures is judged by a target performance comparison. Such a comparison might lead to a reformulation of targets.

We will now once again, and in more detail, describe the phases of substance chain management, by means of a fictitious example. A company produces sports wear and sells it via specialized dealers. Via that same way, a consumer inquiry reaches the firm asking how rain wear made of synthetic fabrics will be disposed of after use, and whether ecological criteria are of relevance in this context. The company has a great number of environmentally aware customers, hence it takes this inquiry seriously. Anyhow, it has already reflected on this question, inspired by the new law on a recycling economy and by its general company philosophy to be always among the ecological pioneers in its sector. As in this situation, furthermore, an important competitor makes publicity with taking back and recycling used garments, there is no choice for the company but become active.

But the company cannot attain its goal alone as it has not yet been concerned with disposal so far. Therefore, the idea as well as the necessity arises to cooperate with other companies along the product chain. Then scopes for action are examined and alternatives are searched for, still within the company, and conditions for their realization are checked. The company orders an institute to ascertain the alternative ways of disposing of the plastics rain wear and find out about ecological advantages and disadvantages of respective procedures. This will possibly lead to changes in the company's production process, the costs of which will have to be assessed. A system of taking garments back will presuppose the preparedness of dealers to provide containers for their collection. The collected garments will have to be transported to a recycling facility, they will have to be recycled into the original product (or into some other useful item). The volume of recycled garments will have to reach a minimum level so as to make the system economically efficient. This might require cooperation with the competitor. That means that a whole series of questions have to be answered in the preparatory phase, which concern ecological meaningfulness, technological realizability, sales prospects in the market, costs, etc., and which are decisive for the success of planned measures. Within the company, these issues are, e.g., dealt with by the environmental department, the departments for product development, for controlling, and for marketing, and these may draw on the support of specialized institutions.

With an already more definite concept now, the firm contacts potential partners in commerce, among competitors, among disposal institutions, forwarding agents, etc., and presents them the idea of a collective system for taking back and recycling the garments. Then negotiations ensue to define goals of cooperation and targets regarding the individual partners. Moreover, tasks are assigned to individual partners, and contractual agreements are made to ensure their realization. So at this stage, the operative translation of agreements into activities is planned, by determining temporal, financial, personal etc. framework conditions for the new tasks and making provisions for a continuous exchange of information.

The next step concerns the implementation of the arrangements made, which, again, takes largely place within individual firms. As has been agreed upon, information is exchanged about development and success of the cooperation, for instance with respect to customers' acceptance, the volume of garments collected during a certain period, their quality, actual costs, etc. At a certain moment determined during the initial stage, success is evaluated by means of an interim stock-taking. Have goals been attained, which the individual companies had formulated for their cooperation? Is cooperation to be continued in the same or in an altered way?

Conclusion

Companies that strive for ecological optimization along the product chain will profitably base their efforts on information instruments like eco-balance, EMAS, eco-controlling, or ISO 14001. Of essential relevance for successful substance chain management projects are, apart from ecological information, a series of economic data. If a company changes its accounting system to a material and energy flow cost accounting this will be of help for assigning costs for environmental protection according to their causes and thus for identifying further optimization potentials. Finally, it is quite sensible to extend the scope of 'classical' quality control by including ecological criteria or integrating environmental and quality management. So, substance chain management does not so much require the development of new instruments as rather the adaptation and integration of existing ones.

References

Ankele, K. (1996): Vom einzelbetrieblichen Umweltmanagement zum Stoffstrommanagement, in: Ökologisches Wirtschaften, 5/1996, ökom, Munich.

Enquête-Kommission "Schutz des Menschen und der Umwelt" (1993): Verantwortung für die Zukunft. Wege zum nachhaltigen Umgang mit Stoff- und Materialströmen. Economica, Bonn.

Fichter, K. (ed.) (1995): Die EG-Öko-Audit-Verordnung. Mit Öko-Controlling zum zertifizierten Umweltmanagementsystem, Hanser, Munich.

de Man, R.; Ankele, K.; Claus, F.; Fichter, K.; Völkle, E. (1997): Aufgaben des betrieblichen und betriebsübergreifenden Stoffstrommanagements, UBA text volume 11/97.

Schramm, E.; Buchert, M.; Bunke, D.; Lehman, S.; Reifenhäuser, I.; Steinfeldt, M.; Strubel, M.; Weller, I.; Zundel, S. (1996): Stoffflüsse ausgewählter relevanter chemischer Stoffe: Produktliniencontrolling, UBA text volume 80/96.

11 Sustainable Development – From Guiding Principle to Industrial Tool at the Hoechst Company

Frank Ebinger, Christoph Ewen, Christian Hochfeld

The Hoechst Company has committed itself to the guiding principle of sustainable development. The Group's own sustainability and that of society as a whole thus gain high priority. This is a commitment that Hoechst needs to reconcile with its orientation to short-term profit expectations. A field is thus outlined that can contain serious inherent contradictions between the long-term and short-term perspectives.

Hoechst commissioned the Öko-Institut (Institute for Applied Ecology) to identify paths by which the guiding principle of sustainable development can find entry in the strategy and management of the Hoechst Group. The objective: in addition to an ethical orientation, Hoechst needs a strategic radar for the developments taking place in the market and in the societal environment – in order that medium – and long-term perspectives can take their place beside short-term expectations.

While the Strategic Management Process (SMP) already gives Hoechst a tool for the derivation of commercially based strategic options for action, there has previously been a lack of such a tool for the issues connected to sustainable development.

With this in mind, the Öko-Institut developed a product assessment tool (PROSA – Product Sustainability Assessment). The present report describes this tool, and gives a resume of the results of its exemplary application. Moreover, it states how Hoechst can use the tool effectively to implement the principle of sustainable development in corporate practice.

What Requirements Does the Guiding Principle Pose?

The principle of sustainable development aims at resolving the two interacting global crises: the overstepping of natural limits (environment, resources, risks), and insufficient development opportunities in large parts of the world (inadequate satisfaction of elementary human needs). Sustainability goes beyond purely ecological objectives and embraces socio-economic concerns: The focus is placed not only on production, nor only on the individual product; the issue is the contribution of Hoechst, with its activities, to the solution of these two crises.

Products that contribute to sustainable development must thus fulfil two requirements: They must contribute to satisfying elementary needs, and they must minimize the associated environmental burdens.

One and the same product can be assessed differently if it is used for different applications. Thus the same product (sorbic acid) will be assessed as being more sustainable when used for the preservation of essential foodstuffs than when it is used for shower gels. While in the case of certain foods the loss through spoilage and thus also the demands placed on the environment can be reduced, shower gels lead to increased demands on the environment compared to soap – without any elementary human needs being additionally satisfied.

If attention shifts to the final use of the products, and thus to the customers of the customers, then the consideration of sustainability leads to a stronger demand-side orientation: Decisive indications of future market developments ensue – also and particularly for a globally operating group.

Requisites for a "Sustainable Hoechst"

In order to implement the guiding principle of sustainable development in the activities of the Group, Hoechst needs:

1. A corporate policy statement: The guiding principle of sustainable development must be made clear and tangible to all staff. It must be possible to formulate it in a short and memorable form, and to translate it into quantifiable objectives.
2. A tool: The implementation of the corporate philosophy requires appropriate tools. In Phase I of the cooperation between Hoechst and the Öko-Institut, a plan of action was developed which in Phase II was then applied to two representative products and further refined. The core of this second phase of the project, to which the present report refers, is a product assessment tool set in the context of sustainable development.
3. Optimized structures: A new guiding principle interferes with conventional principles. If it is to become efficacious, its implementation must be promoted. This involves addressing the boundary conditions effective within the corporation and in its environment – particularly the restructuring currently taking place at Hoechst.

The Tool PROSA – Product Sustainability Assessment

As a guiding principle for society, sustainability cannot be simply expressed in numbers. It does lead to concrete directions for action – but not in the sense that an exactly quantifiable target is stipulated and is then implemented according to a definite schedule. What is called for is rather a searching and learning process, that reveals and normatively substantiates the connections between the products and the guiding principle of sustainability. The task of PROSA is to structure this process, and to transpose it into proposals for action.

In order to assess connections to the guiding principle of sustainability, the tool refocuses attention away from the individual product, and towards the needs behind it. Attention must be directed not only to the direct customer of Hoechst, but also to the consumers behind the customer. The social and ecological systems in which the product and its application are situated need to be analyzed. The distance between Hoechst's direct product and the final consumer is shorter for pharmaceutical and agricultural products, and longer for industrial chemicals. Nonetheless, the connection can be made for the latter, too.

The ecological and social connections crucial to the assessment are largely situated outside of Hoechst's gates. The integration of these external connections into the process is thus an essential condition for PROSA to be successful. To structure the connections, the Öko-Institut has defined four levels. As the "system pyramid" illustrates, each successive level extends the perspective of the previous one:

- the product level (Hoechst's product inclusive of upstream chains)
- the level of the product line (product inclusive of further processing and distribution)
- the application level (functional use of the final product)
- the level of the area of need (reference to consumers; principal areas of need are e.g.: health; food; housing; mobility; recreation; education)

PROSA can address all of the four levels of the pyramid. Its purpose is to identify operative and strategic options for action for Hoechst, proceeding from product assessment. It is crucial to the understanding of the assessment process that it delivers relative rankings. These rankings result from comparison with alternatives – at each level of the system pyramid.

The Five Steps of PROSA

PROSA comprises five steps. Proceeding from a system analysis, product assessment indicators derived from Agenda 21 and the Rio follow-up process are selected. These include fundamental preconditions to sustainability, which, if violated, can lead to termination of the process.

The application of the indicators leads to an assessment in comparison with product alternatives. System analysis and indicator selection are thus sensitive steps, and should be underpinned by bringing in outside expertise.

The Steps of PROSA:

1. System analysis
2. Sustainability references and indicator selection
3. Indicator application for product assessement
4. Analysis of influencing factors
5. Derivation of options for action

The assessment is visualized by positioning the product in a "Sustainability Portfolio". These combine the two axes of "environmental effects" and "development contribution". Through this positioning, this step of PROSA already delivers strategic indications. While a ranking in (the best) Quadrant IV indicates an expansion of the business, a ranking in (the worst) Quadrant I suggests – unless optimizations are to be expected – a termination of the business. Where this is the case, it is not the most purposeful solution to sell off the business; this would not change the real situation. It is more important and more promising to develop alternatives and to convert production processes.

Products and their applications can be optimized along two axes – the axes represent different regions of the world: While in the highly industrialized countries above all ecological optimization (eco-efficiency) is called for, for the so-called developing countries the contribution to development is more important (to meet unmet needs).

Strategic decisions should be oriented to the expected future: The fourth and fifth steps of PROSA have the function of estimating the course that those societal developments will take which are of relevance to the product and to the underlying area of need. How will, for instance, the area of agriculture and food in China develop? Here sustainable and less sustainable, likely and less likely scenarios can be drawn up. Within the framework set by these conceivable developments, the market opportunities of optimized or new products can be estimated. The understanding here is this: Even if thoroughgoing sustainable development is improbable within the next few years, products with enhanced sustainability can offer major market opportunities. Strategic options for action can then be formulated on the basis of these considerations.

Testing the Tool in Practice

In order to tailor the tool to practice, it was applied during its development process in two representative Business Units (now group companies). Two products – sorbic acid and Trevira Spunbond – were assessed with regard to their sustainability in two regions – China and Germany. While data availability and time constraints prohibited a full assessment of the application of sorbic acid in China, the assessment delivered first strategic options for action for Spunbond Trevira.

The findings in brief:

Sorbic acid is a chemical preservative that is mainly used for food preservation, but also for cosmetics and pharmaceuticals, and recently for water lacquer. In addition to the product as such, which can be compared to other chemical preservatives, the "application" level (using the example of sliced bread) and the "area of need" level (using the example of food) were examined.

In this connection, Agenda 21 refers to: combating poverty; changing consumption patterns; promoting sustainable agriculture. While in China the main concern is the adequate supply of the population with hygienically irreproachable food as a part of combating poverty, the concern in Germany is the development of ecologically acceptable production and consumption patterns.

At the product level, the direct toxicological comparison with other food preservatives reveals distinct benefits for sorbic acid. At the application level, the consideration of the example of sliced bread shows that the use of sorbic acid is "worthwhile" in terms of resource conservation if more than about three percent of the total volume of sliced bread that is produced is reliably prevented by sorbic acid from spoiling. At the "area of need" level, considering the range of foodstuffs currently preserved with sorbic acid it can be said for Germany that sorbic acid makes no noteworthy contribution to sustainable development. For China this may be different – the available data permitted no assessment of this.

Trevira Spunbond is a technical textile manufactured from the polyester PET. Its main application is currently as a supporting material for bituminous roofing sheets (Spunbond/ModBit), which are used to seal flat roofs. The ecological comparison of a conventional 2-layer bitumen sheet roof sealing with the high-polymer sheets used alternatively in Germany gives a better result for Spunbond/ModBit than for PVC sheets, but a slightly worse result than for high-polymer sheets made of chlorine-free plastics. In Germany today flat roofs are almost exclusively built for industrial and commercial buildings, so that there is no direct connection to basic human needs. From a sustainability perspective, there is no demand for additional industrial and commercial buildings in Germany, and thus no indirect connection either. For an improvement of the sustainability of Trevira Spunbond, the outcome of the assessment for Germany and the other western markets is that the focus should initially be placed on improving its environmental characteristics.

In China, 3 layers of ragfelt are still almost exclusively used for flat roof sealings, compared to which a single-layer Spunbond/ModBit sealing is ecologically better and further provides a qualitative leap with regard to service life and impermeability. However, it is as yet only used for high-value buildings. Residential buildings are typically built with flat roofs in China. Due to the higher initial investment costs and a disregard for total cost accounting over the useful life of buildings, Spunbond/ModBit is not yet used in this sector. In China a high demand for housing continues to prevail, and the average living space per capita is small. There is thus a very high potential for satisfying basic needs – however this is not yet effectively harnessed for Spunbond. It may be expedient to influence the structure of the downstream value-added chain by means of cooperative partnerships, in order to remove the barriers identified in this sector (actor cooperation, product system cooperation).

Optimizing the Use of the Tool

While the operative level will mainly be able to derive environmental improvements from the use of PROSA, the strategic level can use the tool to align its product portfolio to sustainability criteria.

The Board of Management of the strategic management holding company (SMH) has the following alternatives:

The implementation of PROSA is

1. stipulated in binding form for all operative units,
2. recommended, parallel to this, Sustainability Projects are started and network initiated in central business areas,
3. recommended in non-binding form.

The experience made with the pilot application of PROSA shows that the group companies are in principle willing and motivated to cooperate, but that a successful application of the tool is difficult without support and control by the strategic level. Decentralization harbours the danger that superordinate and strategic guiding principles are not actually implemented in corporate practice. This is illustrated by the experience made at Monsanto, which is also working on the operative implementation of the guiding principle of sustainable development.

On the other hand, the increased devolution of responsibility to the group companies resulting from Hoechst's decentralization is entirely as the sustainability principle would suggest – decentralization, devolution of responsibility and participation are important elements of sustainable development.

The Öko-Institut thus proposes that Alternative 2 be selected. This makes fruitful the benefits of decentralization, without the disadvantages of non-binding and arbitrary implementation. If the guiding principle of sustainable development is to have more than proclamatory value for Hoechst, then the SMH Board must accompany and support the process. To do so, the following measures appear appropriate:

The products of different group companies in part address the same areas of need – their system pyramids overlap at the base. A cooperation among companies with this common feature does not come about of its own accord. Here the SMH should support the group companies in applying the tool by initiating networks that cut across and bring together different business areas. Such networks have the further benefit of generally reinforcing the capacity of the overall group to act according to group principles, which may otherwise suffer due to decentralization.

The SMH should identify central business areas that promise growth in which it demands the results of PROSA parallel to the results of the SMP.

It is the very essence of the tool that it analyses and includes connections extending beyond the gates of the company. Here outside actors are important. Particularly when analysing social influencing factors, it is recommendable to involve outside experts in order to avoid the danger of setting up a system that is incapable of grasping connections beyond its own orbit, and in order to ensure the function of the system as a sensitive early warning system. For Hoechst's credibility, too, it will be important that the process is reviewable from the outside. It is thus purposeful to involve outside representatives of social groups. The SMH should set up guidelines for this.

Hoechst should participate in its political environment in the debate on framework conditions for development towards sustainability. Through this, Hoechst not only does justice to its social responsibility, but improves the market opportunities of its – in future more sustainable – products.

Outlook

The terms of reference of Sustainable Hoechst are now open for scrutiny. In the view of the Öko-Institut, it is an unresolved issue whether and to what extent Hoechst will take the path towards sustainability.

The sustainable realignment of business activities – ranging from changes in the product portfolio to taking a public stand for sustainable framework conditions – is a process that will extend over years. It cannot be assessed at the present point whether Hoechst will resolutely pursue this path.

Past patterns of behaviour suggest that Hoechst will only partly adopt the proposed steps towards sustainability. The Öko-Institut considers inadequate the intention expressed by Hoechst, with reference to the reshuffled organizational structure, to only recommend the sustainability instrument PROSA in a non-binding form to the group companies. A valid assessment of Hoechst's progress can certainly only take place in one or two years from now.

The Öko-Institut will closely monitor developments, and will publish an evaluation of the progress made towards Sustainable Hoechst by the end of 1998 at the latest. It will further, within its (modest) means, seek to influence the stakeholders relevant to Hoechst (investors, top management, staff and customers) such that the Sustainable Hoechst project is realized.

As the next milestones on this path, the Öko-Institut recommends:

In consultation with the group companies concerned, the SMH Board should identify a central area of need (or a part of such an area) for which a Hoechst-wide "Sustainability Project" is initiated.

For an international group, a strategic radar only then makes sense if it extends to the social developments on the most important international markets. When broadly applying PROSA, care should thus be taken that the practice and cultures of other regions are also considered (in particular Asia and North America).

In addition to product assessment, Hoechst should tackle other fields, too: These include e.g. organizational and human resources development, the possible shedding of jobs, financial management, research and development and the sphere of policy affairs, which need to be oriented appropriately.

Final note

At first sight it may appear dubious whether, in times of globalization and shareholder value, the orientation of a globally operating industrial group towards more sustainability is really possible. Nonetheless, the tool presented here does make it possible to identify points of conflict with the principle, and to recognize and work towards win-win situations.

For Hoechst, sustainability means that the Group prepares itself for the future and at the same time makes the future. Sustainable Hoechst can serve as a corporate philosophy that gives the Group cohesion through shared social objectives beside its quest for maximum returns.

A holding company that withdraws to the confines of financial management, bases its portfolio decisions solely on short-term corporate rationality, and otherwise leaves it up to the operative units to handle the guiding principle of sustainable development for themselves will not be able to implement this principle – and neither internally nor in the eyes of the public will it be able to credibly represent it.

References

Öko-Institut (Hrsg.) (1997): Hoechst nachhaltig – Sustainable Development : Vom Leitbild zum Werkzeug, Darmstadt.

C

International Organizations and Networks on Environmental Management

International Organizations and Networks Promoting Environmental Management

As sections A and B of this book show, concepts and practical experience of other countries provide valuable impulses for one's own work in companies, authorities, or science. To support the international exchange of experience and information even beyond this book, we will list below international organizations and networks which foster environmental management and efforts of companies to do sustainable business. National organizations or networks are included only insofar as they are part of an international network. The list does not claim to be complete. The same holds for the international publications and a number of Internet homepages concerning environmental management and sustainable ways of companies doing business, which are listed at the end of this section.

The 'Institut für ökologische Wirtschaftsforschung' (Ecological Economics Research Institute IÖW, non-profit private limited company) deals with international developments in environmental management and keeps contact with many international organizations. For further information, please turn to:

Institut für ökologische Wirtschaftsforschung (IÖW)

Giesebrechtstr. 13
D-10629 Berlin
Tel.: 030 884 594-0
Fax: 030 882 54 39
E-Mail: mailbox@ioew.b.eunet.de
Contact: Klaus Fichter

1   Asian Productivity Organization

Founded

in 1961

Members

Asian governments which are members of the "Economic and Social Commission for Asia and Pacific" (ESCAP) of the United Nations.

Tasks and Goals

The Asian Productivity Organization aims at increasing economic productivity in Asia by means of cooperation, so as to ensure a balanced social and economic development. The Asian Productivity Organization is concerned with:

- propagating information that helps to improve productivity,
- looking after Asian regions,
- building up institutes.

The Asian Productivity Organization concentrates on the following economic sectors: industry, services, and agriculture.

Activities, Projects

The Asian Productivity Organization is mainly active with regard to qualification measures in the following fields:

- quality management,
- information technology,
- environmental management,
- sustainable development in agriculture.

Each year, the Asian Productivity Organization holds a number of workshops concerning environmental management.

Address

Asian Productivity Organization
8-4-14, Akasaka,
Tokyo, 107 Japan
Tel.: +81 3 3408 7221
Fax: +81 3 3408 7220
E-Mail: apo@gol.com.

2   Environmental Auditing Research Group

Founded

in 1991

Members

136 persons from Japanese universities, authorities, auditing companies, and private companies, as well as 23 Japanese firms.

Tasks and Goals

The Japanese Environmental Auditing Research Group (EARG) aims at actively contributing to developing an environment-friendly society in Japan by interdisciplinary promotion and propagation of adequate practices for environmental auditing. The EARG fosters the exchange of knowledge and experience of persons from different occupational fields and sectors at a national level and furthermore is engaged in promoting international exchange concerning environmental audit and environmental management.

Activities, Projects

- regular meetings and workshops,
- studies concerning the relevance of environmental audits,
- projects with respect to special subjects: "Checklists for Environmental Auditing", "Environmental Reports", "Environmental Management Systems of Local Governments".

Publications

- monthly reports and a quarterly newsletter (in Japanese) for members,
- several books, published only in Japanese, though: "Introduction to Environmental Auditing" (1992), "Key Concepts of Environmental Auditing" (1994), "An Easily Understandable Introduction to Environmental Auditing and Environmental Management" (1995), "Benchmarking of Environmental Reporting" (1996, an English summary of this book can be ordered).

Address and Contact

The Environmental Auditing Research Group – EARG Office
6f Dai-2Daitetsu-Bldg.
1-4-11 Shinjuku
Shinjuku-ku
Tokyo 160, Japan
Tel.: +81 3 3353 3788
Fax: +81 3 3353 3757

3   European Business Council for a Sustainable Energy Future – e⁵

Founded

in February 1996, as an association.

Members

Members are companies, associations, and institutes, among them:

- AEG household appliances (Germany),
- DanfossA/S (Danmark),
- ENRON Europe Ltd. (Great Britain),
- INTEGeR.consult (The Netherlands),
- SUN WATT (France),
- Solel Consumer N.V. (Belgium),
- EUROSOLAR (Europe),
- World Fuel Cell Council, registered society (Europe),
- CE DELFT,
- EUROPEAN SMALL HYDROPOWER ASSOCIATION,
- VERBAND DER DEUTSCHEN ENERGIEMANAGER (association of German energy managers),
- WUPPERTAL INSTITUT (Germany),
- WWF Germany.

Tasks and Goals

e⁵ aims at organizing energy demand and supply as well as mobility in economy and society in a way that efficiency and prosperity are maintained without causing irreparable environmental damages and long-term economic problems in Europe or elsewhere. The association is concerned with:

- using the potential of new technologies in the fields of energy, construction, and traffic,
- taking seriously scientific insights about the greenhouse effect,
- supporting a climate protection policy that has to be market-efficient and has to consider external costs,
- using the *least cost planning potential*,
- creating incentives for various actors to contribute to climate protection,
- promoting market introduction of technologies for renewable energies in industrialized and developing countries soon,
- reducing subsidies that are not climate-compatible,
- fostering technology transfer from the North to the South.

Activities, Projects

Press conferences, meetings, and activities in the fields of:

- using the potential of integrated resource planning,
- developing integrated concepts for energy-optimized house systems, as, for instance, solar architecture, heat insulation, heating, lighting, and household appliances,
- having the large potentials of energy efficiency, co-generation of heat and power, and renewable energies taken into account in UN climate protection protocols,
- promoting a liberalization of the EU energy market in favour of energy efficiency on the demand side, co-generation of heat and power, and renewable energies.

Publications

- e^5 -ffects – Newsletter,
- e^5 Homepage: http://www.e5.org,
- booklet representing the association.

Address and Contact

European Business Council for a Sustainable Energy Future – e^5
c/o GERMANWATCH
Berliner Platz 23
53111 Bonn
Tel.: 0228/60492-0
Fax: 0228/6049219
E-Mail: Gemanwatch.Bn@Bonn.Comlink.Apc.Org
e^5 Homepage: http://www.e5.org
Contact: Ms Barbara Assheuer

4   European Environmental Reporting Scheme

Founded

In 1996 for the first time, the European Environment Award for good environmental reporting on the part of companies and organizations was awarded.

Organizors

The European Environment Award is awarded collectively by the national auditor associations ACCA (GB), Royal NlvRA (The Netherlands), and FSR (Danmark). In the years to come, further national auditor associations and scientific institutions like the Ecological Economics Research Institute IÖW in Berlin are to join them.

Tasks and Goals

This award is meant to:

- acknowledge efforts of individual companies and organizations with regard to environmental reporting that is technically adequate and suited to target groups,
- make excellent company environmental reports nationally and internationally known,
- create incentives for companies and organizations to take into account the increased requirement for information on the part of customers, investors, authorities, and social groups,
- motivate more companies and organizations to report about their environmental performance and the environmental impacts of their activities.

Activities, Projects

The organizors of the European Environment Award for good environmental reporting evaluate and rank environmental reports in their respective countries. So, the British auditor association ACCA has awarded a National Environment Award for good environmental reports since 1991. Since 1994, the German Ecological Economics Research Institute IÖW in cooperation with the Förderkreis Umwelt future e.V. regularly evaluates environmental reports and statements of German companies.

Environmental reports ranking high at the national level participate in a contest for the European Environment Award. The respective evaluation is done by an independent jury of experts from different European countries.

Addresses and Contacts

ACCA – Association of Certified Chartered Accountants
29 Lincoln's Inn Fields
London WC2 A 3EE
Tel.: +44 171 242 6855
Fax: +44 171 831 8054
E-Mail: adamsr@acca.co.uk
Contact: Mr Roger Adams

Deloitte & Touche
HC Andersens Blvd. 2
DK-1780 Copenhagen
Tel.: +45 3376 3333
Fax: +45 3376 3940
Contact: Preben Sorensen

Deloitte & Touche
Orlyplein 50
NL-1043 DP Amsterdam
Tel.: +31 20 606 11 00
Fax: +31 20 681 19 87
Contact: Johan Piet

Institut für ökologische Wirtschaftsforschung (IÖW)
Giesebrechtstr. 13
D-10629 Berlin
Tel.: 030 884 594-0
Fax: 030 882 54 39
E-Mail: mailbox@ioew.b.eunet.de
Contact: Klaus Fichter

5   European Partners for the Environment

Founded

in 1993

Members

The European Partners for the Environment (EPE) see themselves as "network of networks". Members are individual European national and local authorities, companies, trade unions, research institutes, environmental associations, and non-governmental organizations, as well as individual persons from these institutions.

Tasks and Goals

EPE is a European "strategic alliance" of different social actors. It aims at contributing to sustainable, environment-friendly development by improving the dialogue and cooperating beyond sectorial and institutional borders. EPE does not understand itself as a lobbying organization but as a new broad coalition striving for a constructive dialogue and collective measures irrespective of its members' partly different interests.

Activities, Projects

Conferences, publications and creation of networks through information measures. Intensified studies of, among others, the following subjects:

- environmental management systems,
- principles of sustainable development in the fields of: transport, tourism, agriculture, nutrition, water, housing construction,
- influence of company-internal environment protection on working conditions,
- free globel trade and environment.

Publications

- EPE InfoService,
- EPE Sourcebook of Partnership Initiatives, 1995–1996,
- Information about EPE: http://www.EPE.be.

Address and Contact

European Partners for the Environment
Ste op Gelrode 123
Rotselaar
B-3110, Belgium
Tel.: +32 16 581 391
Fax: +32 16 581 391
Internet: http://www.EPE.be

6   European Roundtable on Cleaner Production

Founded

in 1994

Members

The European Roundtable on Cleaner Production (ERCP) is a largely informal network of persons from companies, authorities, universities, research institutes, institutions for further education, and non-governmental organizations. There is no formal membership. An executive committee has been existing since 1996.

Tasks and Goals

The ERCP aims at supporting sustainable, environment-friendly development by fostering international dialogue and cooperation concerning environment-friendly products, processes, and services. The ERCP understands "cleaner production" as a "strategy to continuously improve eco-efficiency". This is to be attained by:

- avoiding and reducing wastes,
- reducing energy and water consumption, use of polluting raw materials and processes, as well as non-renewable resources,
- reducing pollution during the whole product life cycle,
- reducing pollution in all social fields.

The ERCP offers a European platform for stimulation, development, and propagation of initiatives to improve eco-efficiency. It focusses on environment-friendly production processes and products in industry, agriculture, fishery, forestry, and tourism.

Activities

- since 1994, an annual international conference,
- workshops.

Publications

- Internet: www.teknologisk.no
- application papers for speeches at the annual conferences.

Address and Contact

European Roundtable on Cleaner Production
National Institute of Technology
24C., P.O. Box 2608 St. Hanshaugen,
N-0131 Oslo
Fax: +47 22 11 12 03
Secretariat:
Nina Norton
Tel.: +47 22 86 51 89
E-Mail: norn@teknologisk.no

7   Global Environment Management Initiative – GEMI

Founded

in 1990

Members

The Global Environment Management Initiative (GEMI) is a union of 22 North American big companies that together employ more than one million persons and achieve a turnover of more than 400 billion US dollars. GEMI North America has two sister organizations in Mexico (INICIATIVA GEMI) and Venezuela (Venezolana Para el Ambiente, IEVA).

Tasks and Goals

GEMI strives for being the global leader in propagating company strategies by means of which top performance in the fields of safety, environment and health protection can be reached. GEMI aims at:

- improving companies' performance as to environment protection and occupational safety by providing practical assistance,
- creating a flexible, market-oriented, autonomous environment protection model which improves the cost-benefit relation,
- improving the compatibility of environmental and safety goals with business conditions.

It is GEMI's motto that "companies help companies to achieve top performance in the fields of security, environment protection, and occupational safety".

Activities, Projects

- annual conferences,
- working groups,
- producing publications and documents regarding environmental management,
- being informed about developments in the field of international standardization of enviromental management.

Publications

- Environmental, Health and Safety Training: A Primer (guidelines to introduce environmental and safety measures in a company, with practical examples),
- Environmental Reporting and Third Party Statements, results of a study,
- Environmental Self-Assessment Program (ESAP),
- GEMI Conference Proceedings (since 1993),
- ISO 14001 Environmental Management System Self-Assessment Checklist,
- Incentives, Disincentives, Environmental Performance and Accountability for the 21st Century, IDEA 21,
- Work Group Reports.
- Homepage: http://www.GEMI.ORG

Address and Contact

Global Environment Management Initiative
1090 Vermont Avenue, NW
Third Floor
Washington, DC 20005 USA
Tel.: +1 202 296 7449
Fax: +1 202 296 7442
Homepage: http://www.GEMI.ORG
Contact: Tammy Marshall

8   Green Cross International

Founded

in 1993 in Japan, under the direction of Mikhail Gorbachev.

Members

Green Cross International (GCI) comprises 18 organizations in different countries. Priority of its work is cooperation with other organizations, companies, associations, universities, and governments.

Tasks and Goals

Green Cross International is an international network focussing on environment and sustainable development. GCI aims at a worldwide change of values and awareness, which it hopes to achieve by multifarious reflection and international cooperation.

Activities, Projects

- peer programme: avoiding accidents in connection with chemical transports, by means of increased attention.
- promoting a sustainable tourism: CGI projects are supposed to show that sustainability and prosperity for a country with respect to tourism are not mutually exclusive.
- professional environmental training: projects in Bolivia, Czechia, Estonia, and Kenya are supposed to better distribute technical know-how, skills, and technologies, by joining national and international experts.
- promoting national and international laws to protect the earth.
- demilitarization: projects concerned with militarization as one of the main causes for environmental destruction are supposed to foster demilitarization.
- environmental education: certain projects are supposed to foster environmental awareness in private households.
- trade and sustainable growth.

    More information via: http://greencross.unige.ch/

Publications

- self-representation
- http://greencross.unige.ch/

Address and Contact

Green Cross International
PO Box 80
1231 Conches – Geneva, Switzerland
Tel.: +41 22 789 1662
Fax: +41 22 789 1695
E-Mail: gci@unige.ch
http://greencross.unige.ch/

9   Greening of Industry Network

Founded

in 1991

Members

The Greening of Industry Network (GIN) is an international network for research and policy development and includes more than 1300 persons from science, authorities, companies, economic associations, trade unions, and non-governmental organizations. GIN is directed by a 20-person council and has coordination offices in Europe (university Twente, Enschede, The Netherlands) and North America (Clark University, Worcester, USA).

Tasks and Goals

The Greening of Industry Network aims at stimulating, coordinating, and promoting important research projects about a sustainable economy, and ensuring high-quality research. In this way, the network wants to contribute to a sustainable, environment-friendly society. It is the task of the annual conferences to bring researchers and practitioners together so as to enable a discussion about long-term strategies for sustainable development and measures for their implementation.

Activities, Projects

There are three fields of activities in GIN:

- annual international conferences (since 1991).
- publications: apart from publishing conference proceedings and books, GIN is engaged in publishing the international review "Business Strategy and the Environment", in cooperation with John Wiley & Sons, Great Britain.
- establishing and maintaining a communication network.

Publications

- conference proceedings.
- GIN's quarterly newsletter appears in the review "Business Strategy and the Environment" of John Wiley & Sons, Great Britain.
- "Environmental Strategies for Industry – International Perspectives on Research Needs and Policy Implications" – Island Press 1993.
- "The Greening of Industry Resource Guide and Bibliography" – Island Press 1996.
- "The Greening of Industry for a Sustainable Future: Building an International Research Agenda".

To be ordered at Island Press: Fax: 707 983 6414, California, USA.

Addresses and Contacts

The Greening of Industry Network
European Coordinators: Ms Ellis Brand and Mr Theo de Bruijn
Center for Clean Technology and Environmental Policy CSTM
University of Twente
PO Box 217
NL-7500 AE Enschede
Tel.: +31 53 489 3215
Fax: +31 53 489 4850
E-Mail: e.m.l.brand@cstm.utwente.nl

U.S. Coordinator: Kurt Fischer
The George Perkins Marsh Institute
Clark University
950 Main Street
USA – Worcester, Massachusetts 01610-1477
Tel.: +1 508 751 4607
Fax: +1 508 751 4600
E-Mail: kfischer@vax.clarku.edu

10   International Network for Environmental Management e.V. – INEM

Founded

activities started in 1989, formal act of foundation in 1991.

Members

The International Network for Environmental Management (INEM) comprises 23 national associations for environmentally aware company management, nine "Cleaner Production Centers" and eight organization committees for the foundation of national associations for environmentally aware company management in 34 countries of all five continents. INEM includes, via its more than 23 member associations, about 5000 companies.

Tasks and Goals

INEM aims at globally fostering, developing, propagating and applying principles and methods of environmentally aware management and at founding as well as supporting national business associations for environmentally aware management all over the world. Environmentally aware management, transcending the field of environmental engineering, concerns all company areas, so, for instance, material management, product development, marketing, personnel management, and company strategy. Among the tasks that INEM has set itself are the following:

- reducing industrially produced pollution,
- propagating information about economic successes due to active environment protection,
- fostering a preventive and integrated environment protection instead of a restorative one,
- supporting environment-protective technologies and environment-friendly products, production processes, and services,
- supporting the implementation of environmental guidelines in companies,
- assisting especially small and medium-sized companies,
- assisting companies in developing and threshold countries.

Activities, Projects

INEM anually holds an international industry conference concerning sustainable development. Moreover, INEM initiates numerous projects, like, for instance:

- gathering methods for small and medium-sized companies to introduce an environment-oriented management system according to the EU regulation on environmental management and environmental audit (1838/93),
- gathering case studies concerning environment-oriented management, e.g. in Central and Eastern Europe,

- collecting statistic data on distribution of and obstacles to environment-oriented management in Poland, Czechia, and Hungary (Global Environmental Management Survey),
- carrying out several projects and workshops concerning environmentally aware management in Poland, Czechia, and Hungary, and concerning
- world religions and environment-oriented company management.

Publications

- INEM bulletin.
- conference proceedings (in English and German):
- "Integrated Environmental Management for Business in Hungary",
- "World Religions and Environment-Oriented Company Management",
- "Case studies in Environmental Management in Small- and Medium-Sized Enterprises", vol.1 (collection of 13 practical case studies, in English),
- Winter (ed.): The Environmentally Aware Company (10 languages).

Addresses and Contacts

INEM e.V.
Bahnhofstraße 36
D-22880 Wedel (Holstein)
Tel: 0 41 03/8 40 19
Fax: 0 41 03/1 36 99
E-Mail: Inem@on-line.de
Home page: http://www.inem.org

ECO-BALTIC Secretariat
Osterstraße 58
D-20259 Hamburg
Tel: 040/49 07-404
Fax: 040/49 07-401
E-Mail: Eco-baltic@on-line.de

National Associations for Environmental Management (members of INEM):

Australia
Environment Management Industry Association of Australia (EMIAA)
GPO Box 2231
Brisbane QLD 4001
Tel.: +61 7 3229 8522
Fax: +61 7 3229 8577
E-Mail: emiaa@emiaa.org.au
Homepage: http://www.emiaa.org.au
Contact: Mr John R. Cole

Austria
Bundesweiter Arbeitskreis für umweltbewußtes Management (BAUM)
Aspettenstrasse 48
A-2380 Perchtoldsdorf
Tel.: +43 1 86632 0
Fax: +43 1 86632 33
E-Mail: gutwinski@cybertron.at
Contact: Mr Hans Werner Autz

Brasil
Sociedade para de Incentivo e Apoio ao Gerenciamento Ambiental (SIGA – INEM Brasil)
Rua do Russel, 300
Apartment 401
22210-010 Rio de Janeiro, RJ
Tel.: +55 21 205 5103
Fax: +55 21 205 4386
Contact: Mr Amauri Solon Ribeiro

Canada
Canadian Chamber of Commerce (CCC)
1160-55 Metcalfe
Ottawa, ON K1P 6N4
Tel.: +1 613 238 4000
Fax: +1 613 238 7643
Contact: Mr Tim Reid

China

National Information Center of Environmental Science and Technology (NICEST)
A Back Yard Constructional Ministry
Bai Wanzhuang
100835 Beijing
Tel.: +86 10 6839 3845
Fax: +86 10 6839 3245
Contact: Mr Zhao Feng

Colombia

Promoción de la Pequeña Empresa Eco-Eficiente en Latinoamérica (PROPEL)
Carrera 12 No. 93–31
Oficina 406
Bogotá
Tel.: +57 1 622 1314
Fax:.+57 1 622 1247
E-Mail: propel@colomsat.net.co
Contact: Mr Carlos H. Barragan

Czechia

Czech Environmental Management Center (CEMC)
Jevanska 12
100 00 Prague 10
Tel.: +420 2 628 0957
Fax: +420 2 7758 69
E-Mail: cemc@ecn.cz
Contact: Mr Roman Vyhnánek

Danmark

Erhvervslivets Ledelsesforum for Miljofremme (ELM Danmark)
203 Klampenborgvej
DK-2800 Lyngby
Tel.: +45 4587 5030
Fax: +45 4587 3216
Contact: Mr Frede Bjerg Petersen

France
Partenariat Entreprises Collectivés Environnement (Orée – INEM France)
42, rue du Faubourg Poissonnière
F-75010 Paris
Tel.: +33 1 4824 0400
Fax: +33 1 4824 0863
E-Mail: oree@hol.fr
Home page: http://www.oree.com
Contact: Mr Philippe Marzolf

Germany
Bundesdeutscher Arbeitskreis für umweltbewußtes Management (BAUM-INEM Germany)
Tinsdaler Kirchenweg 211
D-22559 Hamburg
Tel.: +49 40 810 101
Fax: +49 40 810 126
E-Mail: info@baumev.de
Home page: http://www.baumev.de
Contact: Mr Maximilian Gege

Hungary
Környezettudatos Vállalatirányítási Egyesület (KOVET-INEM Hungária)
Munkácsy Mihály u. 16
H-1603 Budapest
Tel.: +36 1 131 6763
Fax: +36 1 33 207 87
E-Mail: gergelytoth@mail.neti.hu
Contact: Mr Gergely Tóth

Ireland
Irish Productivity Centre (IPC)
42–47 Lower Mount Street
Dublin 2
Tel.: +353 1 662 3233
Fax: +353 1 662 3300
E-Mail: ipc@indigo.ie
Contact: Mr Norbert Gallagher

Israel
Society of Industry for Ecology (ALVA)
c/o Ormat Industries Ltd.
P.O. Box 68
70650 Yavne
Tel.: +972 8 9433 732
Fax: +972 8 9439 901
Contact: Mr Michael Gill

Lithuania
Inzinerines Ekologijos Asociacija
(IEA – INEM Lithuania)
a/d 560, Dowinikonu 4
2001 Vilnius
Tel. and Fax: +370 2 22 3879
Contact: Mr Rimantas Budrys

Malaysia
Environmental Management & Research Association of Malaysia (ENSEARCH)
30A Jln SS21/58, Damansara Utama
47400 Petaling Jaya
Selanger Darul Ehsan
Tel.: +60 3 717 7588
Fax: +60 3 717 7596
E-Mail: jenny@cetec.po.my
Contact: Mr C.K. John

Mozambique
Forum Empresarial Para o Meio-Ambiente (FEMA)
c/o Cartonagens de Mocambique
451-A, OUA´s Avenue
P.O. Box 180
Maputo
Tel.: +258 1 401 362
Fax: +258 1 401 622
Contact: Mr Joáo Viseu

Philippines
Philippine Business for the Environment (PBE)
3/F DAP Building
San Miguel Ave.
1601 Pasig City
Metro Manila
Tel.: +63 2 635 3670
Fax: +63 2 631 5714
Contact: Ms Grace Favile

Slovak Republic
Association of Industrial Ecology in Slovakia (ASPEK)
Drienova 24
826 03 Bratislava
Slovak Republic
Tel.: +421 7 230 833
Fax: +421 7 5797 546
Contact: Mr Andrej Soltes

Slovenia
Drustvo Poslovodnih Delavcev Slovenije (DREVO)
Dunajska 106
61113 Ljubljana   ·
Tel.: +386 61 168 1169
Fax: +386 61 168 3106
Contact: Ms Polona Blagus Smonig

Sweden
Näringslivets Miljöchefer (NMC)
Fleminggatan 7
Box 8133
S-104 20 Stockholm
Tel.: +46 8 6571 075
Fax: +46 8 6533 193
Contact: Mr Olle Blidholm

Switzerland
Schweizerische Vereinigung für ökologisch bewußte Unternehmensführung (ÖBU)
Im Stieg 7
CH-8134 Adliswil
Tel.: +41 1 709 0980
Fax: +41 1 709 0981
E-Mail: 106160.266@compuserve.com
Contact: Mr Arthur Braunschweig

Zimbabwe
Environmental Forum of Zimbabwe (EFZ)
P.O. Box AC 270
Ascot
Bulawayo
Tel.: +263 9 700 13
Fax: +263 9 796 03
Contact: Mr James W. Harrower

Cleaner Production Centers (members in INEM):

Cleaner Production Centers:
Since 1992, donors from USA and Norway as well as various organizations like
the United Nations, UNEP (United Nations Environmental Programme) and
UNIDO (United Nations Industrial Development Organizations), have fostered
the spreading of environment-technical know-how (emission reduction, waste
minimization, etc.) by supporting Cleaner Production Centers.

Cleaner Production Centers aim at supporting industry and trade in their
efforts for a more sustainable way of doing business with simultaneously saving money, increasing productivity and reducing pollution.

Australia
Australia Centre for Cleaner Production (ACCP)
Building 251
RMIT Bundoora East
Plenty Road
Bundoora
Victoria 3083
Tel.: +61 3 9407 6060
Fax: +61 3 9407 6061
Contact: Mr Allen W. Morley

Czechia
Czech Cleaner Production Center (CCPC)
Politickych veznu 13
111 21 Prague
Tel.: +420 2 260 620
Fax: +420 2 260 639
E-Mail:czechcpc@czn.cz
Contact: Mr Vladimir Dobes

Estonia
Pollution Prevention Center at the Estonian Management Institute
21 Sutiste Street
EE 0034 Tallinn
Tel.: +372 2 521 629
Fax: +372 6 411 882
E-Mail: anne@emi.estnet.ee
Contact: Ms Anne Randmer

India
National Cleaner Production Centre (NCPC)
5–6, Institutional Area
Lodi Road
110003 New Delhi
Tel.: +91 11 461 1243
Fax: +91 11 461 5002
Contact: Mr S.P. Chandak

Latvia
Latvian Pollution Prevention Center (LPPC-INEM Latvia)
Perses Street 2–518
LV-1011 Riga
Tel.: +371 7828 250
Fax: +371 7828 251
E-Mail: natalia@lppc.org.lv
Contact: Ms Natalia Ladoutko

Lithuania
Pollution Prevention Center at the Institute of Environmental Engineering (APINI)
K. Donelaicio 20–307
LT-3000 Kaunas
Tel.: +370 7 224 655
Fax: +370 7 209 372
E-Mail: ppc.kaunas@apini.ktu.lt
Contact: Mr Jurgis Staniskis

Slovak Republic
Slovak Cleaner Production Centre (SCPC)
Hanulova 9/A
845 01 Bratislava
Tel.: +421 7 782 681
Fax: +421 7 784 467
E-Mail: feckova@cvt.stuba.sk
Contact: Mr Anton Blazej

Tunisia
Centre de production Plus Propre
(CP3 – INEM Tunisie)
Colisée Saula
Escalier D – 2 ème étage
2092 El Manar
Tel.: +216 1 872 688
Fax: +216 1 870 766
E-Mail: naftir@cp3.tn
Contact: Mr Rachid Nafti

Zimbabwe
Cleaner Production Centre of Zimbabwe (CPCZ)
P.O.Box BW 294
Borrowdale
Harare
Tel.: +263 4 883 865
Fax: +263 4 883 864
E-Mail: cpczim@harare.iafrica.com
Contact: Mr Lewin Mumbemuriwo

INEM Organizing Committees for the Foundation of National Associations for Environmental Management:

Belgium
INEM – Belgium Organizing Committee
Mr Heinz Werner Engel
Communication, Environment, Gestion
246, Chaussee d'Alsemberg
B-1180 Brussel
Tel.: +32 2 346 6970
Fax: +32 2 346 6465
E-Mail: hwengel@arcadis.be
Contact: Mr Garsett Larosse

Ecotopia Virtual Corporation vzw
Antwerpsesteenweg 461
B-2390 Westmalle
Tel.: +32 3 312 9373
Fax: +32 3 309 2874
E-Mail: Garsett@ecotopia.be

Bulgaria
INEM – Bulgaria Organizing Committee
Mr Branimir Natov
Bugarian Industry Association
16–20, Alabin Street
1000 Sofia
Tel.: +359 2 545 066/ +359 2 595 416
Fax: +359 2 872 604
E-Mail: bia@bulmail.sprint.com

Chile
Red Inter Empresarial de Gestion Ambiental
(RIEGA – INEM Chile Organization Committee)
Ms Nicola Borregaard
CIPMA
Avda. Holanda 1109
Casilla 16362
Santiago 9
Tel.: +56 2 334 1096
Fax: +56 2 334 1095
E-Mail: nborrega@cipma.cl

Ghana
INEM – Ghana Organizing Committee
Mr Neustadt Amarteifio
EMPRETEC Ghana Business Advisory
Service
Private Mail Bag, Accra
1st floor SAT Builders, Merchant Building
Kwame N'krumah Avenue
Accra
Tel.: +233 21 668 571
Fax: +233 21 665 574

Pakistan
INEM – Pakistan Organizing Committee
Mr Junaid Ahmad
c/o Federation of Pakistan Chambers of Commerce and Industry (FPCCI)
1st floor, PIDC House
M.T. Khan Road
Karachi
Tel.: +92 21 568 1897
Fax: +92 21 568 9455

Russian Federation
INEM – Central Russia Organizing Committee
Mr Yuri Piskulov
Environmental Committee of the Russian Chamber of Commerce and Industry
Iljinka Str. 6
RUS – 103684 Moscow
Tel.: +7 095 203 4189
Fax: +7 095 203 3586

Mr Aleksey E. Ignatijev
Institute for Natural Resources Management
34 B., Cheryemushkinskaya
RUS – 117259 Moscow
Tel.: +7 095 120 0151
Fax: +7 095 124 9553
E-Mail: mara@iepp.msk.su

INEM – Northwest Russia Organizing Committee
Mr Stanislav Ozjabkin
SOFDEC Ltd.
Box 37
RUS – 198103 St. Petersburg
Tel.: +7 812 218 9654
Fax: +7 812 251 2526 (tel. & fax)
E-Mail: sofdec@itec.spb.su

Ukraine
INEM – Ukraine Organizing Committee
Mr Vladimir Tikhii
Environmental Education and Information Center (EEIC)
2 Skovorody Street
252070 Kiev
Tel.: +380 44 416 6039
Fax: +380 44 274 2417
E-Mail: eeic@gluk.apc.org

11   International Sustainable Development Research Network

Founded

in 1995

Members

about 350 scientists and practitioners from 34 countries.

Tasks and Goals

The International Sustainable Development Research Network (ISDRN) aims at bringing together persons from different occupational fields and countries so as to foster research about sustainable development and propagation of best approaches in practice. The ISDRN is concerned with:

- extending interdisciplinary research in the field of 'sustainable development',
- providing a forum for discussions by means of annual conferences,
- enabling high-quality research by the journal "Sustainable Development",
- exchanging information on important initiatives,
- stimulating education initiatives in the field of 'environment and sustainable development',
- cooperating with other networks, institutions, and non-governmental organizations so as to support strategies in the field of 'sustainable development'.

Activities, Projects

- annual conferences,
- research groups and workshops,
- "ISDRN Electronic Conference Site", a possibility for members to communicate via e-mail: Address: listserver@hud.ac.uk, Subject: New Subscription, Command: subscribe isdrn- list

Publications

- ISDRN bulletin,
- "Sustainable Development", journal,
- publication of current member list, twice a year,
- information booklet on the annual conferences.
- "ISDRN Electronic Conference Site", a possibility for members to communicate via e-mail: Address: listserver@hud.ac.uk, Subject: New Subscription, Command: subscribe isdrn- list

Address and Contact

International Sustainable Development Research Network
Centre for Corporate Environmental Management – CCEM
University of Huddersfield
Queensgate,
Huddersfield. HD1 3DH
West Yorkshire, UK
Tel.: +44 (0)1484 472262
Fax: +44 (0)1484 472633
E-Mail: ISDRN@hud.ac.uk
Contact: Ms Linda Orwin

12   Network for Environmental Management and Auditing

Founded

in 1993

Members

The Network for Environmental Management and Auditing (NEMA) includes more than 400 persons from science, companies, authorities, and consulting firms, mostly from Great Britain.

Tasks and Goals

NEMA aims at facilitating the exchange of information and experience in the fields of environmental management and environmental auditing so as to improve companies' environmental performance and avoid unnecessary double work in the development of instruments for environmental management and in research. As an interdisciplinary network, NEMA strives for bringing together scientists, practitioners, and politicians.

Activities, Projects

- building up and maintaining an address data base of members and experts,
- holding workshops, usually every two months,
- publishing results of research and workshops,
- supporting information exchange in the fields of environmental management and environmental auditing.

Publications

- "Environmental Management Systems and Cleaner Production", edited by Ruth Hillary,
- NEMA workshop documents,
- NEMA member list.

Address and Contact

Network for Environmental Management and Auditing
Centre for Environmental Technology
Imperial College for Science, Technology and Medicine, University of London
48 Prince's Gardens
London SW7 2PE
Tel.: +44 171 589 5111
Fax: +44 171 581 0245
Contact: Ms Ruth Hillary

13   United Nations Environment Programme – Industry and Environment – UNEP IE

Founded

The centre "Industry and Environment" of the United Nations Environment Programme (UNEP-IE) was founded in 1975.

Organizational Embeddedness

The centre "Industry and Environment" is part of the United Nations Environment Programme (UNEP). UNEP in turn is a subsidiary organization of the United Nations.

Tasks and Goals

UNEP has founded the centre "Industry and Environment" as a means to develop environment-friendly and safe production and consumption models for its member nations. UNEP-IE supports international exchange of information and experience and thereby means to contribute to an improved cooperation and to realizing sustainable ways of economic activities. UNEP-IE is concerned with:

- attaining consensus with regard to preventive environment protection in the economy,
- giving support with respect to policy development and formulation of strategies,
- promoting the integration of ecological criteria into industrial production,
- fostering the exchange of information about environment-friendly technologies and environment-oriented management approaches.

Activities, Projects

UNEP IE carries out numerous projects and activities. Among them are:

- "Query-Response Service": a service to answer questions on the part of industry concerning pollution and legal regulations,
- information pool with topical information about environment-protective measures in the economy worldwide,
- workshops and training for qualification in the field of environment protection,
- pilot projects,
- promotion of Cleaner Production Centers.

Cleaner Production Centers:
Since 1992, donors from USA and Norway as well as various organizations like the United Nations, UNEP (United Nations Environmental Programme) and UNIDO (United Nations Industrial Development Organizations), have fostered the spreading of environment-technical know-how (emission reduction, waste minimization, etc.) by supporting Cleaner Production Centers.

Cleaner Production Centers aim at supporting industry and trade in their efforts for a more sustainable way of doing business with simultaneously saving money, increasing productivity and reducing pollution.

Publications

- publications representing "Cleaner Production":
  publications listed below concerning the subject 'cleaner production' can be ordered via: SMI (Distribution Services) Limited, P.O.Box 119 Stevenage, Hertfordshire SG1 4TP, England – Fax: +44 (1438) 748844.
  Among them are:
  - Eco-Efficiency and Cleaner Production, Charting the Course to Sustainability – a publication of UNEP and WBCSD about the relation between cleaner production and eco-efficiency
  - Cleaner Production in China: A Story of Successful Cooperation, 1996
  - Life Cycle Assessment: What it is and How to do it, 1996
- reviews about technical guidelines, political reports, training documents, and case studies.
- quarterly circulars informing about:
  1. APELL (Awareness and Preparedness for Emergencies at local level): APELL was founded to prevent industrial accidents or to minimize their impacts.
  2. "Cleaner Production",
  3. introduction of environment-friendly technologies, above all in developing countries,
  4. "Ozone Action" reports on company-internal and product-related measures against ozone depletion.
- the quarterly "Industry and Environment" provides the possibility for scientists, practitioners, and persons from politics and administration to exchange practical experience and research results.
- further journals and reports are, e.g.:
  - *CSD Update* reports on national and international activities in the field of sustainable development which are also related to certain chapters of the Agenda 21. More information via: CSD Room DC2-2252, United Nations, New York NY, 10017, Tel.: +1 212 963 3170, Fax: +1 212 963 4260, E-Mail: alvarenz-rivero@un.org

- Environmental Technology Assessment: reports twice a year on the activities of UNEP-IE in the field of evaluating environmental technologies. Further information via: UNEP-IE, Paris, France.
- UNIDO – UNEP National Cleaner Production Center Programme Newsletter (NCPC Newsletter): gives twice a year the most topical information from the centers that work within the NCPC programme. Further information via: UNEP-IE, Paris, France.

Address and Contact

United Nations Environmental Programme – Industry and Environment
39–43. Quai Andre Citroen
75739 Paris Cedex 15 – France
Tel.: +33 (1) 44 37 14 50
Fax: +33 (1) 44 37 14 74
E-Mail: unepie@unep.fr

14   Valdez Society, Japan

Founded

in 1991

Members

150 individual persons, five corporations, and three 'non-profit' organizations. The Valdez Society, Japan, is member of the U.S. American "Coalition for Environmentally Responsible Economies" (CERES).

Tasks and Goals

The Valdez Society, Japan, sees its main task in supporting research groups and studies concerned with issues of company-internal environment protection, environmentally aware consumption, and investments in Japan, and supports publication of research results.

Activities, Projects

Research groups of the Valdez Society study the following issues:

- environment-friendly products,
- "green" investments,
- recycling,
- perspectives of a sustainable society,
- comparison of officially published environmental data in the USA and in Japan.

Publications

"The Valdez Society News", twice a month, in Japanese

Address and Contact

Valdez Society
#302 Nishikawa Bldg.
2–3–7 Kouji-machi
Chiyoda-ku
Tokyo, Japan
Tel.: +81 3 3230 1237
Fax: +81 3 3263 9175
Contact: Ms Kimie Tsunoda

15   Verein für Umweltmanagement in Banken, Sparkassen und Versicherungen – VfU

Founded

in 1994

Members

Banks, savings banks, and insurances in Germany, Austria, and Switzerland.

Tasks and Goals

This first sector-specific association aims at developing new strategies and appropriate instruments for environmental management in banks, savings banks, and insurances and at supporting their practical implementation. Sector-specific workshops are developed that focus on: environmental knowledge, management, communication, organization, and cooperation. A further goal is the transfer of the Environmental Mangement and Audit Scheme EMAS to financial services. In addition, the association is concerned with identifying, structuring, and publishing relevant environmental information for its members.

Activities, Projects

- exchanging, in regular meetings, experience about already achieved progress in environmental management as well as about weak points identified,
- procuring appropriate contacts and qualified experts in case of ecological problems and need for consulting,
- holding workshops for further ecological training,
- holding workshops about environmental aspects in loan business (credit allocation), and in insurance business, and about ecological benchmarking.

Publications

- "Umweltberichterstattung von Finanzdienstleistern" – a guideline concerning contents, structure, and indicators in environmental reports of banks and savings banks.

Address and Contact:

Verein für Umweltmanagement in Banken, Sparkassen und Versicherungen – VfU
Wilhelmstraße 28
53111 Bonn
Tel.: 0228/ 766 8494
Fax: 0228/ 766 8496
E-Mail: 101330.3112@compuserve.com
Contact: Ms Hiltrud Bendels

16   World Business Council for Sustainable Development – WBCSD

Founded

in January 1995

Members

The World Business Council for Sustainable Development (WBCSD) is an association of 120 big international companies, which, due to their size, feel responsible for the environment and for principles of environmental management. The member companies declare their readiness to make their skills, experience, and knowledge in the field of environmental management available to the WBCSD.

Tasks and Goals

The WBCSD aims at improving cooperation among members and with governments or other organizations occupied with environmental mangagement. The WBCSD is concerned with:

- promoting outstandingly efficient environmental management in its member companies,
- fostering environmental awareness of the top management, so as to be able to influence a company's environmental management,
- influencing political developments so as to stimulate companies to start with environmental management,
- carrying out pilot projects to illustrate progresses in environmental management,
- being a global speaker of member companies as far as environmental management is concerned.

Activities, Projects

Various activities focus on the following issues:

- sustainable production and consumption,
- climate and energy,
- trade and environment,
- financial markets,
- natural resources,
- development of concepts to improve the situation of forest clearing and paper industry.

Publications

- "Sustain", a gratuitous quarterly information journal,
- "WBCSD Annual review", published every February
  further publications:
- Eco-Efficient Leadership
- Environmental Assessment
- Trade and Environment
- Eco-Effiency and Cleaner Production
- Sustainable Production and Consumption
- Towards a Sustainable Paper Cycle
- A Changing Future for Paper
- Signals of Change

Address and Contact

World Business Council for Sustainable Development
160, route de Florissant
CH-1231 Conches-Geneva, Schweiz
Tel.: +41-22 839 31 00
Fax: +41-22 839 31 31
E-Mail: info@wbcsd.ch
Internet: www.wbcsd.ch
Contact: Ms Inge Fellay

17   Further International Sources of Information

Specialized Journals

- *The ASIAN Quarterly on Pollution Prevention*
  reports on regional activities to prevent production-related pollution. It focusses, among other things, on the following subjects: LCA, ISO 14000, and successful practical examples. Further information: Ms Julie Haines, US AEP, 1720 Eye Street, Suite 600, Washington DC 20006, Tel.: +1 202 835 0333, Fax: +1 202 496 9720, E-Mail: jhaines@usaep.org
- *At the Source*
  a circular informing about environmental activities in North America, published by the "Great Lakes Pollution Prevention Centre. Further information via: GLPPC, 265 N Front Street, Suite 112, Sarnia, Ontario, N7T 7X1, Tel.: +1 800 667 9790, Fax: +1 519 337 3486, E-Mail: sarnia@glppc.org.
- *Business and the Environment*
  monthly international newsletter focussing on company-internal environment protection and environmental management. Further information: Cutter Information Corporation, 37 Broadway, Suite 1, Arlington, MA 02174-5552, USA, Tel.: +1 617 641 5125, Fax: +1 617 648 1950, Web-Site: www.cutter.com/envibusi/.
- *Business Strategy and the Environment Journal*
  International newsletter focussing on company-internal environment protection, environmental management, and environmental management research. Further information: Wiley & Sons, ERP Environment, P.O. Box 75, Shipley, West Yorkshire, BD17 6EZ, Great Britain, Tel.: +44 1274 53 04 08, Fax: +44 1274 53 04 09.
- *Cleaner Production in Central and Eastern Europe*
  reports on topical developments, publications, and projects in Central and Eastern Europe as to cleaner production. Further information via: EAP Task Force, Non-Member Countries Branch, Environment Directorate, OECD, 2 rue Andre Pascal, 75775 Paris CEDEX 16, France, Fax: +33 01 45 24 96 71.
- *Ecocycle*
  reports twice a year about instruments of eco-balancing, management, and product policy. More information via: Environment Canada, Hazardous Waste Branch, Ottawa, Ontario, Canada K1A 0H3, Tel.: +1 819 997 3060, Fax: +1 819 953 6881, E-Mail: kbrady@synapse.net (http://www.doe.ca/ecocycle)
- *Eco-Management and Auditing*
  international newsletter focussing on environmental management and environmental management research. Further information: Wiley & Sons, ERP Environment, P.O. Box 75, Shipley, West Yorkshire, BD17 6EZ, Great Britain, Tel.: +44 1274 53 04 08, Fax: +44 1274 53 04 09.

- *Environmental Accounting and Auditing Reporter*
  monthly international newsletter concerning eco-balancing of companies, environmental cost accounting, environmental reporting, and environmental auditing. Further information: Milan Pau, IBC Publishing, Gilmoora House, 57–61 Mortimer Street, London W1N 8JX, Great Britain, Tel.: +44 171 673 43 83, Fax: +44 171 636 64 14
- *Green Design*
  reports quarterly about the relation between sustainable development and product design. More information via: SDA, 4560 Mariette, Montreal Quebec H4B 2G2, Tel.: +1 514 482 5033, Fax: +1 514 482 6823, E-Mail: sda@grndsn.login.qc.ca
- *Green Product Design*
  circular about environmentally aware product development. More information via: Green Product Design, Jaffalaan 9, 2628 BX, Delft, The Netherlands, Tel.: (3115) 782 738, Fax: (3115) 782 956, E-Mail: vos@io.tudelft.nl
- *Journal of Cleaner Technology and Environmental Sciences*
  published by the International Association for Clean Technology (IACT), reports quarterly about pollution prevention and environment-friendly technologies. Further information via: IACT Secretariat, Rechte Wienzeile 29/3, A 1040 Vienna, Austria, Tel.: (431) 567 487, Fax: (431) 314 182.
- *Greener Management International – GMI, Journal*
  The GMI Journal addresses in the first place companies interested in environmental management and is concerned with strategies and the operative implementation of environment protection into companies. Further information via: Greener Management International Greenleaf Publishing. 8–10 Broomhall Road Sheffield S10 2DR, UK., Mr John Stuart, Publication Co-ordinator, Tel.: +44 114 266 37 89, Fax: +44 114 267 94 03
- *Ökologisches Wirtschaften*
  newsletter every two months concerning ecological economics research and sustainable development. Published by the 'Institut für ökologische Wirtschaftsforschung (IÖW) gGmbH' and the 'Vereinigung ökologischer Wirtschaftsforschung (VÖW) e.V.' Further information: IÖW, Giesebrechtstr. 13, D-10629 Berlin, Tel.: +49(0)30 884 594-0, Fax: +49(0)30 882 54 39, E-Mail: mailbox@ioew.b.eunet.de.
- *Sustainable Development*
  newsletter concerning sustainable development in companies, communities, and the Third World. Further information: Wiley & Sons, ERP Environment, P.O. Box 75, Shipley, West Yorkshire, BD17 6EZ, Great Britain, Tel.: +44 1274 53 04 08, Fax: +44 1274 53 04 09.
- *UNEP Industry and the Environment Review*
  quarterly newsletter addressing environmental experts in industry, research, and administration. Published by the United Nations Environment Programme – Industry and Environment, 39–43. Quai Andre Citroen, 75739 Paris Cedex 15, France, Tel.: +33(1)44 37 14 50, Fax: +33(1)44 37 14 74, E-Mail: unepie@unep.fr

Homepage Addresses

- Centre for Sustainable Design – http://www.cfsd.org.uk
- Cleaner Production in Central and Eastern Europe – http://www.oecd.org/env/eap
- Ecologica E-TIP Database – http://ecologia.nier.org/index/html
- FAO Homepage – http://www.fao.org
- Institute for Local Self-Reliance (ILSR) – http://gopher.great-lakes.net:2200/0/partners/ ILSR/ILSRhome.html
- International Institute for Sustainable Development – http://www.iisd.ca/linkages/
- International Cleaner Production Information Clearinghouse (ICPIC) – http://www.unepie.org
- International Environment Technology Centre (UNEP IETC) – http://www/unwp.or.jp/ietc/Databases/index/html
- International Hotel Association – http://www.hospitalitynet.nl/iha
- International Institute for Sustainable Development (IISD) – http://iisd1.iisd.ca/
- International Register of Potentially Toxic Chemicals (IRPTC) – http://irptc.unep.ch/pops/, http://irptc.unep.ch/irptc/
- PRenventive Environmental Protection AppRoaches in Europe (PREPARE) – http://cleantechnology.rtc-cork.ie/prepare
- Resource Renewal Institute (RRI) – http://www.rri.org
- Regional Environmental Centre – http://www.rec.hu/
- Sustainable Product Development – UNEP Working Group – http://unep.frw.uva.nl

Autors

Kathrin Ankele ...
born in 1964, masters of Biology, since 1993 researcher at the Ecological Economics Research Institute (Institut für ökologische Wirtschaftsforschung (IÖW) in Berlin). She is working in the research field "Environmental Management" with focus on the following subjects: Life Cycle Assessment of Products, Eco-Balance of Companies, Information Instruments within and between Companies (Material Flow Management), Ecological Evaluation Methods, Implementation of Environmental Management Systems.

Michael Aucott ...
has been involved with environmental issues for over 25 years. He received a master's degree in environmental sciences from Rutgers University in 1987, and is presently enrolled in the Ph.D. program there, researching global chlorine issues. He has been with the New Jersey Department of Environmental Protection since 1987. He has worked on chlorofluorocarbon management and mercury emissions. Presently, he is a research scientist with the Department's Office of Pollution Prevention, where he manages the collection of pollution prevention information and analyzes industrial chemical release, transfer, throughput, and pollution prevention data.

Matteo Bartolomeo ...
Matteo Bartolomeo has got a degree in Economics, he is European Master in Environmental Management and has got a diploma at the International Program on Management of Sustainability, The Netherlands. As PhD He is specialising in the field of environmental performance indicators in industry. As a project manager at Fondazione Eni Enrico Mattei he is working in the field industry and the environment by co-ordinating several research projects on the issue. He is co-founder and director of Avanzi, a newly established research institute dealing with sustainability issues. He is member of the Expert Working Group on Environmental Accounting of the UNCTAD-International Standards on Accounting and Reporting and of the European Environmental Management Association. He is also member of the Environmental Accounting and Auditing Reporter and of the Social and Environmental Accounting Newsletter Editorial Board.

Corinne Boone ...

Ms. Corinne Boone is currently on a one year assignment as Senior Advisor: Sustainable Development with Electricite de France on the E7 Initiative. Her major responsibilities are advising on matters relating to sustainable development in the electricity sector and assisting with the coordination of the E7 Initiative.

Prior to this Ms. Boone was an Advisor, Full Cost Accounting in the Environment and Sustainable Development Division. Her major responsibilities included identification and monetization of externalities associated with generating stations and transmission lines. Her main professional interests include developing strategic and technical methods for integrating sustainable development principles into business decision-making processes.

Ms. Boone has a Bachelor of Arts in Economics and History and a Masters of Environmental Studies (MES) – Integrated Resource Planning and Environmental Economics.

Jens Clausen ...

born in 1958, is a diploma'd engineer, and since 1991 has been doing research as a member of the Ecological Economics Research Institute (IÖW) in the research field "Environmental management". He is manager of the IÖW project office in Hanover. From 1984 to 1991, before he joined the IÖW, he worked in the research and development department of the 'Continental' corporation in Hanover. He is member of the subcommittee 'Environmental Management Systems' of the German Standards Organization (DIN). His activities focus on environmental performance indicators, environmental reporting, and eco-design.

Thomas Dyllick ...

is professor for business administration with a focus on environmental management at the University of St.Gallen, Switzerland. He is chief executive director of the institute for economy and ecology at this university (Institut für Ökologie und Wirtschaft – Hochschule St. Gallen IWÖ – HSG) and chairman of the Swiss association for environmentally aware company management (Schweizerische Vereinigung für ökologisch bewußte Unternehmensführung Ö.B.U.). Prof. Dyllick is member of the Swiss national committee associated with the programme "Human dimensions of global change", member of the board of the association of environmental verifiers, auditors, commissioners and environmental management advisors (Vereinigung für Umweltgutachter, -auditoren, -beauftragte und Umweltmanagementberater VUG) and member of the board of the committee 'Environmental Management Systems' of the Swiss commission on auditing and certifying (Schweizerischer Ausschuß für Prüfen und Zertifizieren SAPUZ).

Frank Ebinger ...
studied business administration at the Universities Fulda and Mainz. From 1995 untill 1996 he worked as a scientific assistant at the Eropean Business School, Oestrich-Winkel. Since 1996 he is a member of the chemistry division at the Öko-Institute in Freiburg. Main focus of his current work are the fields of Sustainable Development (Implementation into industrial practice), Cost-Management, Environmental Management-Systems and Eco-Auditing.

Christoph Ewen ...
studied civil engineering with the main subjects regional and environmental planning at the Technical University of Darmstadt. From 1985 until 1989 he worked as a Research associate of the chemistry division at the Öko-Institute Darmstadt. From 1989 until 1993 he was the Coordinator of Chemistry Division. Since 1993 he is the Deputy Secretary of the Institute and since 1996 he is also Scientific Coordinator of the Institute.

Klaus Fichter ...
born in 1962, has a doctor's degree in economic science from the University of Oldenburg. Since 1993 he has been researcher at the Ecological Economics Research Institute IÖW, Berlin. He is head of the environmental management research department and has been responsible for several pilot projects concerned with implementing the Environmental Management and Audit Scheme EMAS. In 1998 he finished his dissertation on environmental communication and competitivness. His work focusses on: corporate environmental reporting, environmental performance and competitivness, environmental cost accounting, ecologically oriented personnel development, international developments of environmental management.

Gianreto Gamboni ...
studied business administration and economics at the University of Basel (lic. rer. pol.). Postgraduate studies in environmental sciences at the University of Zurich. He worked as a Swiss Bank Corporation financial analyst from 1988 to 1996 and has headed the Environmental Performance Analysis group in the Institutional Asset Management division since 1996.

David M.W.N. Hitchens ...
is professor and head of the department of economics at the Queens University Belfast. He was born in 1948 and held positions at The City University of London, the National Institute of Economic and Social research in London and the Northern Ireland Economic Research Center. He is specialized in questions of productivity and competitivness and did research in numerous projects on this subject. Recently he is active in projects on environmental protection and competitivness and on the interrelationsship of economic and environmental performance.

Christian Hochfeld ...

studied environmental engeneering at the Technical University of Berlin from 1989 until 1996. In 1993 he spent a year for practical training in several countries of Latin America and at the Öko-Institute in Darmstadt in the chemistry division where he did his diploma thesis, too. From 1996 to 1998 he worked as the assistant of the scientific coordinator. Since 1998 he is a member of the chemistry division in Darmstadt. Main focus of his current work are the fields of Sustainable Development (Implementation into industrial practice), Material Flow Management and Green Financing.

Martin Houldin ...

is 44 years old and is working as an environmental assessor for the United Kingdom Accreditation Service (UKAS). He is founder member of the Environmental Management Advisory Group (EMAG) which specialises in advising business on the successful implementation of environmental management systems. Several companies have been guided through BS 7750, EMAS and ISO 14001 (DIS).

From 1975 to 1981 he worked for Ford Motor Company in various positions including production, product development, finance, information systems, internal audit. From 1981 to 1984 he was a management accountant including internal audit with J P Morgan. 1984 to 1988 he worked with KPMG in the field of management consultancy to the financial services industry. He established the environmental unit of KPMG in the UK, and was responsible for a range of environmental management studies with companies in different sectors from 1989 to 1994. He was involved in external auditing.

Mr. Houldin is chairman of the American Chamber of Commerce (UK) Environment Committee, member of the judging panel for the ACCA Environmental Reporting Awards, member of the EARA Training Sub-Committee and member of the environmental committees of the Chartered Institute of Management Accountants (CIMA), the 100 Group of finance directors, the Federation des Experts Comptables Europeens (FEE). Furthermore he is a member of the Institute of Environmental Management (IEM), Environmental Auditors Registration Association (EARA), Chartered Institute of Management Accountants (CIMA), Institute of Management Consultants (IMC) and the Institute of Management (IOM).

Helen Howes ...

Ms. Howes is currently working as Senior Advisor, Environmental Responsibility and Leadership Department in the Corporate Strategies group at Ontario Hydro. Her major responsibilities include supporting the Vice-President in establishing the policy framework and corporate environmental management system to enable the corporation to meet its environmental responsibilities; contributing to the achievement of Ontario Hydro's business targets through the provision of advice on pending and future environmental risks and opportunities; and seeking opportunities for environmental leadership throught the development of strategic partnerships and innovative solutions. Her main professional interests include mechanisms to integrate environment into business decision-making, environmental management systems and practices.

Ms. Howes has a B.Sc (Hon.) Biology from Queens University, Kingston, Ontario; Masters in Environmental Studies from York University, Toronto, Ontario; and a field course in Tropical Ecology (Belize) at Brock University, St. Catharines, Ontario.

Ali Khan ...

Mr. Ali Khan is currently working as a Senior Corporate Financial Analyst in the Financial Strategy and Policy Division of Ontario Hydro. Prior to this, he worked as an Advisor for Full Cost Accounting in the Environment and Sustainable Development Division where he contributed in the development of tools for integrating environmental considerations into business decisions, such as full cost accounting, life-cycle assessment, eco-efficiency, and resource planning. Mr. Khan has also held management and engineering positions with Exxon Chemical and Schlumberger in South Asia, Europe, the Far East and Australia.

Mr. Khan is a member of the LEAD Program of the Rockefeller Foundation. He has earned a Master of Business Administration from the University of Texas at Austin and a Bachlor of Engineering from the NED University of Engineering and Technology in Karachi, Pakistan.

Tron Kleivane ...

born in 1956, is married, has one child and is living in Oslo. He studied in Lilleström and Paris and graduated at the Institut d'Etudes Polititques de Paris in 1984. He was a member of the European Movement in France and an adviser in foreign political questions. In 1985 Tron Kleivane founded the AXESS Information Services AS and was the director until 1989.

From 1989 until 1991 he served as the Director of Strategy at ESSELTE A/S with responsibility for business development within electronic information services. He is the founder and chairman of the European environmental foundation of the European Green Table. Since 1991 he is the executive adviser on strategic issues within the Confederation of Norwegian Business and Industry (CNBI) with special responsibilty for CNBI's EC-strategy, environmental strategy and public opinion strategy.

Hugo Kuijjer ...

started his career at the Inspectorate of Environmental Protection in the province of Noord-Holland after finishing a Higher Technical College qualification in environmental management and ecology in 1975. He was involved in all subjects connected with the tasks of the inspectorate. For example an act on the subject of burying/undertaking, the Watersupply Act, the Spatial Planning Act, and more well known environmental regulations like the Nuisance Act and the Environmental Protection Act.

In 1987 he temporary switched his job (job rotation) and worked for more than half a year as a technical environmental advisor of the Council of State. He advised upon a wide range of appeals – for instance technical advice on permits for stockfarming, a brewery, an opera company, and others. From 1988 till 1990 he was the leader of a project within the Division Nuisance Act of the Ministry of the Environment.

From 1990 he has worked for the Industry-Division of the Ministry of the Environment. He is responsible for the coordination and policymaking of environmental management. Under his guidance a programme (worth 50 million guilders) on the development and the implementation of environmental management systems has been carried out. He is now also involved in the development of a new policy subject: the incorperation of product improvement into environmental management systems.

Thomas Loew ...

is economist and working with the Ecological Economics Research Institute in Berlin since 1994. His activities focus on environmental performance indicators, environmental cost accounting and environmental reporting.

José A. Lutzenberger ...

was born in 1929 in Brazil, as a son of German immigrants, and studied agronomy in Porto Allegre. For some years, he worked in fertilizer firms in Rio Grande do Sul, then he came to BASF, Ludwigshafen, Germany, where he worked in the agronomic services division. Later, he was sent for some years to Caracas and Morocco. Then, in 1971, he experienced a kind of enlightenment, from a Saulus changed into a Paulus, quit his job at BASF and devoted his abilities entirely to free-lance ecological farming. He initiated and participated in model projects which attracted worldwide interest. Furthermore, he acted as an advisor to pulp production plants, abattoirs, tanneries, plants that process soybeans, and any kind of waste and waste water processing facilities. In 1988, in addition to numerous honours, he was awarded the alternative Nobel prize in Sweden and became known worldwide.

From 1990 to 1992 he was appointed Minister of Environmental Affairs of the Brazilian government. Among other things, he contributed to the decision not to built a nuclear bomb, he was engaged in bringing about the Antarctica Convention, and actively participated in organizing the Rio Summit in 1992. One of his most serious concerns is the protection of the Amazonas rain forests.

Among his numerous publications are:

- Lutzenberger (1994): Knowledge and Wisdom must come back together, Folkuniversitetet, Sweden.
- Lutzenberger: We can´t improve nature, in print, Edition Siegfried Pater, Postfach 150106, 53040 Bonn, Tel.: +49-228-236484
- a book about Lutzenberger is: Pater, Siegfried: José Lutzenberger, Das Grüne Gewissen Brasiliens (Brazil's green conscience), Lamuv, Göttingen, Lamuv pocket book Nr. 163, 1994

Maite Mathes ...

was born in 1960 in Bilbao/Euskadi and grew up in Ingelheim at the river Rhine. She worked several years as an animal nurse, studied veterinary medicine in Hanover and wrote her doctoral thesis on breeding problems in small populations of endangered pig breeds. The "Working Group of Critical Veterinarians" has been her spiritual home since 1983 and she started to be involved in ecological farming and ecological economics. Since 1994 she is member of the "Network for Precaring Economy", an interdisciplinary group of women developing a feminist concept for a sustainable economy. Presently she is working at the Humbold University in Berlin to prepare and guide master studies in sustainable development.

Susanne Nisius ...

(nee Grotz) was born in 1968. After completing a degree in Business Administration at the University of Tübingen in 1995, she joined the regional office Heidelberg of the Institute of Ecological Economic Research (IÖW) working in the field of product policy. Her main specialisms in research are in the application of Life Cycle Assessments in companies and their role in business decision making processes.

Frans Oosterhuis ...

Frans Oosterhuis is a researcher at the Institute for Environmental Studies (IVM), Vrije Universiteit, Amsterdam, The Netherlands. He is economist and did work in many multidisciplinary projects, e.g. in the field of product policy. He is co-author, together with Frieder Rubik and Gerd Scholl, of "Product Policy in Europe: New Environmental Perspectives" which was published at Kluwer Academic Publishers in 1996.

Margret M. Pierce ...

is 39 years old, married and resides with her husband, her daughter and three stepchildren in Wilmington, Delaware. She studied chemical engineering at the New Jersey Institute of Technology and finished her studies as a bachelor of science. For 15 years she worked for E.I. DuPont de Nemours & Co., Inc. Wilmington. Her fields of work were: process control and plant design, capital investment management, project management and environmental consulting. Presently she is working as a consultant to the Chambers Works plant (located at Deepwater, NJ) concerning pollution prevention and other environmental regulations.

Takis Plagiannakos ...

Mr. Takis Plagiannakos is a Senior Corporate Planner in the Corporate Strategies group at Ontario Hydro. His major responsibilities include undertaking research to address corproate strategic issues, providing advice on strategic planning and environmental accounting and developing corporate performance measures for monitoring progress towards corporate strategic objectives. His main professional interests are in the areas of environmental accounting, strategic planning and corporate performance measures.

Mr. Plagiannakos has a Bachelor Degree in Economics from the University of Athens, Greece,; a Master of Economics from York University, Toronto, Canada; and a Master of Business Administration also from York University.

Federica Ranghieri ...

Federica Ranghieri, graduated in Economic and Social Studies at Bocconi University, Milan. Since January 1995, she is Researcher in Environmental Accounting and Reporting at the Fondazione Eni Enrico Mattei. She is Responsible for the Environmental Reports Monitoring System, leader of the Environmentalal Benchmarking Studies team at ENI Holding and she is also assisting several companies in preparing their environmental reports and has worked for the Italian Oil industry association in the preparation of a sector environmental accounting and reporting system.

Barbara Reuber ...

Ms. Barbara Reuber is a Senior Corporate Planner in the Corproate Strategies group at Ontario Hydro. Her major responsibilities include development of environmental policies and provision of strategic advice on environmental issues. Her main professional interests include approaches for management of air emissions from industrial facilities and analytical modelling of environmental impacts.

Ms. Reuber has a Bachelor of Applied Science degree in Engineering Science from the University of Toronto and attended the University of Toronto's Institute for Environmental Studies where she obtained a Master of Science degree.

Gerd Scholl ...

is a researcher at the Ecological Economics Research Institute (IÖW), regional-office Baden-Württemberg. He studied economics in Göttingen and Bonn and is working with the institute since 1993. His research areas are instruments of product-oriented environmental policy, product labelling, life-cycle assessment (LCA), and eco-services. He has done studies for the European Commission (DG XI, XII), the Federal Environmental Agency (UBA) and the Federal Ministry for Economy (BMWi). At the time being he co-ordinates a research project dealing with new concepts of product use.

Inge Schuhmacher ...

studied Ecology and Business Administration at the University of Lüneburg (Germany). After participating in the IÖW environmental reporting project, she selected the topic of environmental communications and independent assessment of corporate ecological performance for her thesis in environmental management. She has been pursuing this specialization at Swiss Bank Corporation (now UBS) since August 1995. After joining the Environmental Management Services unit she has been working in the Environmental Performance Analysis group in the in the Institutional Asset Management division.

Tomo Shibamiya ...

is 60 years old. He graduated from the Tokyo Institute of Technology and worked for Toshiba for 32 years in the area of management in manufacturing engineering, management of US factories and environmental management in Toshiba Corporation (HQ). While working in Toshiba he was a subcommittee member of the Global Environment Committee in the Confederation of Japanese Industry KEIDANREN and vice chairman of the Environmental Auditing Committee of the Japan Electrical Manufacturers Association. Furthermore he has been a delegate to a working group of the ISO-Technical Committee 207 "Environmental Management" (TC 207, SC 5, WG 1).

Presently Mr. Shibamiya is director and general manager of the Japan Audit and Certification Organisation for Environment (JACO), Tokyo. He is an environmental lead auditor of JACO.

Springer
und
Umwelt

Als internationaler wissenschaftlicher
Verlag sind wir uns unserer besonderen
Verpflichtung der Umwelt gegenüber
bewußt und beziehen umweltorientierte
Grundsätze in Unternehmens-
entscheidungen mit ein. Von unseren
Geschäftspartnern (Druckereien,
Papierfabriken, Verpackungsherstellern
usw.) verlangen wir, daß sie sowohl
beim Herstellungsprozess selbst als
auch beim Einsatz der zur Verwendung
kommenden Materialien ökologische
Gesichtspunkte berücksichtigen.
Das für dieses Buch verwendete Papier
ist aus chlorfrei bzw. chlorarm
hergestelltem Zellstoff gefertigt und im
pH-Wert neutral.

Springer

Printing: Mercedesdruck, Berlin
Binding: Buchbinderei Lüderitz & Bauer, Berlin